Top 20 Essential Skills for ArcGIS® Online

TOP 20

ESSENTIAL SKILLS FOR ARCGIS® ONLINE

Craig Carpenter
Jian Lange
Bern Szukalski

Esri Press
REDLANDS | CALIFORNIA

Esri Press, 380 New York Street, Redlands, California 92373-8100

Printed in the United States of America.

ISBN: 9781589487802
Library of Congress Control Number: 2024931160

For purchasing and distribution options (both domestic and international), please visit esripress.esri.com.

CONTENTS

ACKNOWLEDGMENTS

This book would not have been possible without the Esri Press team. Many thanks to Catherine Ortiz, Stacy Krieg, David Oberman, Victoria Roberts, Mark Henry, Beth Bauler, and Sasha Gallardo for their vision, support, and work on this book.

Thanks to Sarah Bell, Sarah Hanson, Bethany Scott, Tina Skousen, Blake Stearman, and Jeff Swain for their advice, contributions, and perspectives.

INTRODUCTION

Not long ago, when someone made a map, they scratched a rock or painted a cave. As durable as that technique has been, the ability to share a map was limited. In contrast, by the end of the first chapter of this book, you will have created a publicly accessible web map that includes live weather information. By the end of this book, you will have used a deep learning model to analyze a satellite image, repurposed digital content from ArcGIS® Living Atlas of the World, and created a map from code—using your web browser. You don't need to install anything or store any data on your computer.

This book is designed to give you an overall understanding of ArcGIS Online and teach you how to make maps, primarily by working on a bunch of fun projects. No matter where you're coming from—even if you've never made any kind of map before and are worried about diving into a new technology—this book is for you. You'll be guided through each step on the way to reaching your goals. The book provides plenty of images of the ArcGIS Online environment and easy-to-follow instructions showing you exactly how to perform each task, and you won't be burdened with confusing technical jargon.

We're big fans of learning by doing, so that's our approach. We think that you need to be actively making maps to learn mapmaking. You'll be guided, but you'll also have plenty of opportunities to experiment as you go. We want you to experience the joy of creating, analyzing, and sharing what you make.

HOW TO USE THIS BOOK

About this book

Top 20 Essential Skills for ArcGIS Online has been tested for compatibility with the ArcGIS Online February 2024 release. This book is designed for beginners. No prior GIS knowledge or experience is needed. Each chapter uses one or two tutorials to demonstrate the related skill in a hands-on environment and should take about 45 minutes to complete.

Chapter 1 introduces the ArcGIS Online user interface and navigation. If you're new to GIS or ArcGIS Online, you should start with that chapter. After that, you can work through the book sequentially or complete any chapter in any order because each chapter uses a stand-alone dataset. The last two chapters cover how to make apps that showcase maps that were made previously in the book. If you didn't complete those chapters, you can substitute any other map.

Software and licensing requirements

To perform the tutorials in this book, you'll need a web browser to access ArcGIS Online. For details on required user types, roles, and privileges, go to links.esri.com/AGOL20License.

Use existing credentials

If you have a license that provides the required functionality, you can use your sign-in credentials and proceed.

Activate a Student Use license, provided with participating book purchase

The Student Use license, which may be provided with your book purchase, is available to users in the United States, subject to current terms and conditions. This license provides a range of products, but some chapters require additional products that may not be included with this license.

To see a list of products available with this license and steps for activating this license, go to links.esri.com/Top20License.

1. Locate the license code that comes with participating books.
    - Print textbooks may come with a code printed inside the back cover.
    - E-books purchased through participating delivery platforms may come with a license. Participating e-books purchased through VitalSource will be labeled *courseware*. A code is provided by VitalSource after purchase. Visit links.esri.com/BookCode for help locating this code.
2. Use your code to sign up at links.esri.com/TrialCode.

Additional license options can be found at www.esri.com.

Tutorial data

Steps for accessing tutorial data are provided as needed in each chapter.

Tutorial data for this book is covered by a license agreement that stipulates the terms of use. You can review this agreement at links.esri.com/LicenseAgreement.

Settings

Before getting started, you must confirm or adjust the following settings:

1. Go to arcgis.com and sign in to your account.
2. At the upper right of your browser window, click your username and choose My Settings.
3. On the General tab, under Primary Map Viewer, confirm that Map Viewer is selected.
4. Under Units, confirm that US Standard is selected.

Resources and learning

Tips for success

- Learning mapmaking skills is your goal. Read all the text and take your time. Avoid lightly scanning the instructions or clicking without knowing why. Read the explanations.
- The "Take the Next Step" section at the conclusion of each tutorial is optional but recommended. These sections provide additional suggestions, add functionality, and further refine the project you've completed. While the workflow is fresh in your mind, this section is your chance to solidify what you've learned and further develop your final product.
- The "Workflow" section at the end of many chapters provides a simplified version of what you learned. If you've completed the chapter, the workflow should help you repeat the workflow with other data, whenever you need.

ArcGIS Online resources

Visit links.esri.com/ArcGISOnlineResources for a variety of resources:

- Read blogs describing new functionality and workflows provided by Esri's ArcGIS Online team.
- Sign up for a monthly email newsletter.
- Discover features that have been recently added to ArcGIS Online.
- Watch videos that show everything from the basics of getting started to the nuances of advanced analysis.

ArcGIS Online documentation

Visit links.esri.com/ArcGISOnlineDocumentation for Esri documentation and Help topics that can answer many of your questions and provide in-depth details and troubleshooting tips.

Esri Community

Visit links.esri.com/OnlineCommunity to post questions, share ideas, and engage with other ArcGIS Online users.

Esri Academy

Visit esri.com/training to explore a variety of Esri training products, including the following:

- Interactive, self-paced **web courses** explore workflow-based steps through hands-on software exercises.
- **Massive open online courses (MOOCs)** feature video lectures and hands-on exercises that thousands of people take together.
- **Live training seminars** provide technical presentations delivered by Esri experts, who offer tips, demonstrate software capabilities, and answer questions from the public.
- **Instructor-led training courses** taught by Esri instructors use hands-on exercises and feedback in a classroom environment to teach workflow-based skills and emphasize best practices.

Feedback and updates

Feedback, updates, and collaboration happen at Esri Community, the global community of Esri users. Post any questions about this book at links.esri.com/EsriPressCommunity.

Visit the book's web page at links.esri.com/AGOL20.

CHAPTER 1

Exploring ArcGIS® Online basics

Objectives

- Learn what ArcGIS Online is and what you can do with it.
- Use your ArcGIS Online organizational account to make your first online map.
- Tour and familiarize yourself with the user interface.

Introduction

If you have never made a web map (or any kind of map), this is the perfect place to start. In this chapter, you'll make a web map, learn about the software that your ArcGIS Online organizational account gives you access to, and take a quick tour of ArcGIS Online.

ArcGIS Online is the software you'll use throughout this book to create layers, maps, 3D web scenes, web apps, notebooks, and more. It's software-as-a-service (SaaS), which means you sign in with your browser and you're ready to start making maps—no software installation needed. All the layers, maps, and apps you'll create are saved in the ArcGIS cloud, which makes your maps available online.

ArcGIS Online is an integral part of Esri's suite of seamlessly integrated ArcGIS products, including desktop, mobile, and enterprise software. Whenever you need additional functionality, you can learn about other geospatial products that will help you make maps in more ways.

Web-based GIS

An ArcGIS Online organization is a web-based geographic information system (GIS). Businesses, government agencies, and schools license a subscription to ArcGIS Online, set up their own organizations, and manage all their geographic content in a secure, cloud-based environment. Accounts are then created for anyone who needs to use the software. For example, a water department that wants to manage its infrastructure with a GIS would set up an ArcGIS Online organization and provide organizational accounts to managers, engineers, and technicians. An ArcGIS Online organization stores the accounts and data under an umbrella that's easy for the organization to manage. Layers, maps, and other data that need to remain private can be configured so that only people within the organization can view it. Other maps can be shared with other agencies that also have an ArcGIS Online organization, or maps can be shared online with the public.

Different role and user settings further configure an account's ability to access certain functionality and do certain tasks. An organization's administrator manages settings to create maps or other types of content, collaborate within the organization's groups, or publish work to the public.

If you're using an organizational account provided by your school, your account will probably reside within your school's ArcGIS organization. If you're using the student use license that comes with this book, you'll have your own organization and an account within that organization. In that case, you'll be the administrator of your organization and the sole member—you can call yourself a wolf pack of one. Regardless of the organization you're a member of, you'll have access to a cutting-edge GIS that's waiting to be used, so let's get started!

Tutorial 1-1: Start with a map

"I wisely started with a map."
—*J. R. R. Tolkien*

The best way to get to know the software is to make a quick map. You'll make a map that shows US airports and includes a live layer displaying weather watches and warnings from the National Weather Service.

Sign in to ArcGIS Online

You may have credentials that you received from your school or workplace. If you do, in the next step you can enter your username and password. If you don't have an account yet, see the "How to Use This Book" section in the front of the book and review the "Software and Licensing Requirements" section for licensing options.

1. In a web browser, go to **www.arcgis.com** and click Sign In.
2. Enter your ArcGIS Online username and password, and click Sign In.

> **Hint:** Check the Keep me signed in box if you're using a personal computer and want to stay signed in.

ArcGIS login
Username
Password
Keep me signed in
Sign In
Forgot username? or Forgot password?

> *ArcGIS Online is a dynamic, ever-improving website, so your interface may slightly differ from the images you see in this book.*

Open Map Viewer

Now you'll open Map Viewer and start making a map.

1. On the top navigation bar, click Map.

   Welcome to Map Viewer! Right now, there aren't any layers in the map, so you're looking at the default basemap. For much of the work you'll do, you'll use toolbars on the left and the right of the view. The vertical dark toolbar on the left is the *Contents toolbar*, and the vertical light toolbar on the right is the *Settings toolbar*.

2. On the lower left, click the arrows to expand the Contents toolbar.

The Contents toolbar is where you add layers and publish the map. It's where most settings related to the whole map are located.

3. On the lower right, click the arrows to expand the Settings toolbar.

   This toolbar is used to adjust the settings of whichever layer is selected in the Contents toolbar. When you click a layer in the Contents toolbar, settings appear on the Settings toolbar that are relevant to the layer you selected. In other words, clicking a layer gives you context-dependent settings.

Add a group layer from ArcGIS Living Atlas

ArcGIS Living Atlas of the World is a dynamic atlas of ready-to-use layers, basemaps, imagery, maps, scenes, apps, and services. This catalog gives you an easy start to mapmaking, since you don't have to worry about tracking down basic data that you may need for your map—you just add layers from this catalog.

Later in the book, you'll explore ArcGIS Living Atlas in depth, but for now you'll add a group layer to start building a weather map.

1. On the Contents toolbar (the dark toolbar on the left), click Layers.
2. Click Add.

   The Add layer pane by default shows My content, which is content that you own. Don't worry if you don't have any content yet—that's about to change.

3. In the Add layer pane, click the down arrow next to My content and click Living Atlas.

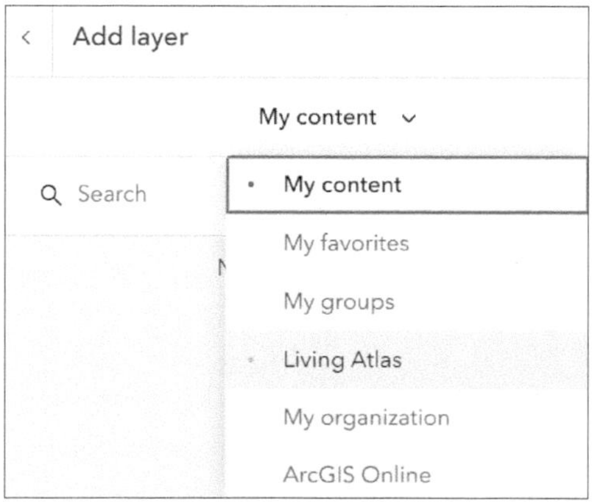

4. Under Living Atlas, in the search box, type **USA Weather Watches and Warnings**.

5. In the search results, locate the layer and click Add.

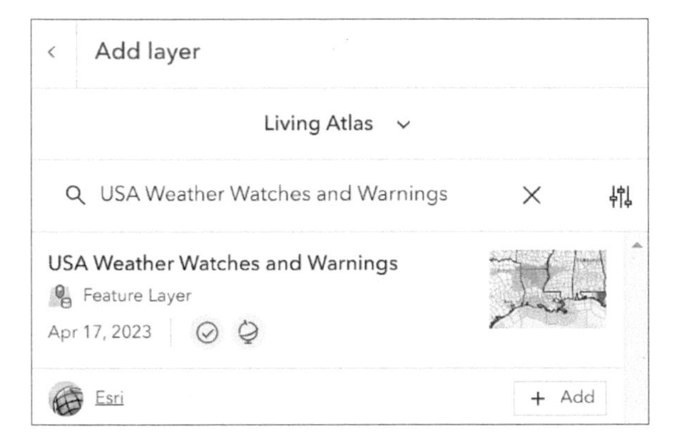

This group layer shows live US weather that's streamed from the National Weather Service.

6. At the top of the Add layer pane, click the back arrow.

7. In the Layers pane, click the arrow to the left of the USA Weather Watches and Warnings group layer.

   Several weather-related layers make up the USA Weather Watches and Warnings group layer. You don't need every layer, so you'll turn one off.

   Layer visibility can be turned on and off by clicking the Visibility button, which appears when you hover over it.

8. Point to the Coastal and Offshore Marine Zones layer and click the Visibility button (eye icon) to turn off the layer visibility.

Style a layer

Now you'll change the style and modify the look of a layer.

1. Click the US States and Territories layer to select it.

   When you selected this layer, the Properties pane on the Settings toolbar opened, containing settings you can modify.

> **Hint:** If the Settings toolbar is not fully expanded, click the arrows on the lower right to expand it.

The Properties pane is organized into several groups of settings that provide an overview of the layer properties.

2. Under Symbology, click Edit layer style.

   On the Settings toolbar, Styles is now highlighted and the Styles pane is active.

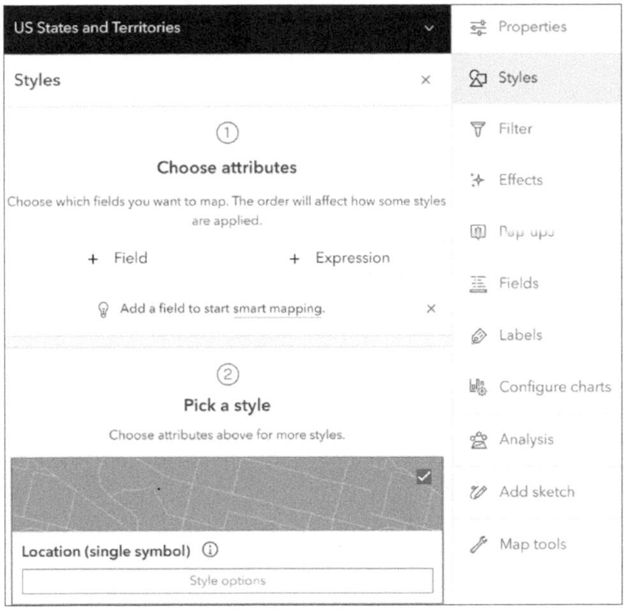

3. Under Pick a style, click Style options.
4. In the Styles pane, under Symbol style, click Edit (pencil).
5. In the Symbol style pane, under Outline color, click Edit.
6. In the Select color pane, choose a dark blue and click Done.

   You've just made a simple change to a style. As you may imagine, there's a lot more you can do to style a layer. You'll get to that later in the book.

7. Close any panes and toolbars to better see the map.

Build from a solid basemap

Basemaps provide background context for the layers you add to your map. Map Viewer has many basemap options from which to choose.

> *Your basemap options will vary depending on your license.*

1. On the Contents toolbar, click Basemap.
2. In the Basemap pane, scroll through the list of basemaps and click some of them to see them.
3. From the list, choose Light Gray Canvas for your final choice of basemap.

Add a layer from ArcGIS Online

You can add any layer that has been publicly shared by anyone with an ArcGIS Online account. These layers can be shared by public agencies, businesses, schools, non-profits, or individuals.

You'll add a layer of US airports, creating a map that can help travelers who may be wondering about weather impacts to an upcoming trip.

1. On the Contents toolbar, click Layers.
2. In the Layers pane, click Add.
3. In the Add layer pane, click the down arrow next to My content and click ArcGIS Online.
4. In the search box, type **owner:Top20EssentialSkillsForArcGISOnline airports**.

   Hint: The first part of this search term, "owner:Top20EssentialSkillsForArcGIS-Online," locates all the public content owned by the account that was created for this book. By adding "airports," you narrow the search to find a specific item with that term. Throughout this book, you'll use this search method to find items.

5. In the search results, locate the USA Airports layer and click it.

   Details about the layer appear. Here, the summary states that the layer was made for this tutorial and is owned by the Top20EssentialSkillsForArcGISOnline account. Under Status, it has been labeled as authoritative, which indicates that it has passed a quality-control check.

6. At the bottom, click Add to map. Close any panes that may be obscuring the map.

   You can see major airports in the context of current weather watches and warnings. You've made a pretty useful map!

7. In the map, click any warnings or watches that may be impacting an airport.

Save your map

Once you save your map, you can share it with your organization, specific groups, or the public.

1. On the Contents toolbar, click Save and open, and click Save as.

When saving content, you need to fill out several fields. This is your opportunity to organize your new content, which may not sound like the most exciting thing, but taking a moment to add a few important details can save you hours of trying to find or organize things.

2. In the Save map window, for Title, type **US Airports and Weather <your initials>.**

> *If you're in a class that's using this book, each student is probably using accounts in the school's ArcGIS organization. Each map (or item) in an organization must be uniquely named, so you need to add your initials to your map name to make it different from the names of the other maps in the classroom. Throughout this book, when you name an item, you'll be expected to add your initials to the end of the name—for example, US Airports and Weather CC. Make sure, of course, that your initials don't match any other student's initials.*

3. For Folder, click the down arrow and choose Create new folder. Type **Top 20 Tutorial Content**.

   You're adding a folder that you'll use to hold all the content you make as you progress through this book. If you happen to be using an account from work, doing so will keep your tutorial activities separate from your work content.

4. For Tags, type **Top 20, ArcGIS Online**. Use a comma to separate tags.
5. For Summary, type **A map of airports and live weather**.
6. Click Save.

   In the upper left of your window, the new title of your map is now visible.

Share your map

1. In the Contents toolbar, click Share map.
2. In the Share pane, click Everyone (public) and click Save.
3. Paste the map URL in an incognito or private window to confirm that you have a publicly available map. If you can see the map, you know it's been published.

   In this tutorial, you took two publicly available layers and combined them into a great first map. You can see how easy it is to make a web map!

Tutorial 1-2: Tour ArcGIS Online in detail

In the first tutorial, you explored Map Viewer, which is where you made your airport and weather map. In this tutorial, you'll take a quick tour of ArcGIS Online and learn how the layers and map you used are stored and managed within the larger system.

Sign in to ArcGIS Online

1. If you are signed out, go to **www.arcgis.com** and click Sign In.
2. Enter your ArcGIS Online username and password, and click Sign In.

   At the top is the main navigation bar.

3. Click Content.

   The blue navigation bar lets you choose from a variety of content locations. You'll explore each in turn.

Tour My Content

Each item you own is available on the My content tab on the blue navigation bar. The map you made in tutorial 1-1 was saved to the Top 20 Tutorial Content folder. You'll open this folder and locate the map.

1. At the far left, under Folders, click Top 20 Tutorial Content.

   Top 20 Tutorial Content

2. Find the map you made in tutorial 1-1.

   US Airports and Weather | Web Map | Feb 7, 2024 | View details ...

   The globe indicates that the item is being shared publicly.

3. On the far right, click Options and click Add to favorites.

Tour My Favorites

My Favorites is where you can access content that you've set as favorite content. This can be content that you or others own, such as content from ArcGIS Living Atlas.

1. On the blue navigation bar, click My favorites.

   As you may have guessed, your web map is here. Using the favorites functionality is a great way to track content and make it easy to find.

Tour My Groups

1. On the blue navigation bar, click My groups.

   A group can be used to share items among people who are creating layers or maps without making those items public. When the work is complete, items can be updated for sharing publicly. You'll search for a group that was created for this book and join it.

2. On the top navigation bar, click the search tool (magnifying glass).
3. In the search box, type **Top 20 Essential Skills for ArcGIS Online** and press Enter.

   On the left, notice that Content is highlighted. Most of the time, you'll be searching for content, so this setting is the default.

4. Click Groups.

   You have filtered your search to find groups. The search is also filtered to show only those groups in your ArcGIS Online organization.

5. Under Filters, turn off the toggle next to the name of your organization.
6. On the group search results for Top 20 Essential Skills for ArcGIS Online, click View details.
7. On the group overview page, in the upper right, click Join this group.

   Just like that, you're in!

8. On the blue navigation bar, click Content.

   Content is shown that has been shared with the group. You'll learn more about accessing and managing items that have been shared this way later in the book.

9. On the blue navigation bar, click Members.

   All the group members are listed. You can collaborate with all of them.

Tour My Organization

1. On the top navigation bar, click Content. On the blue navigation bar, click My organization.

   My organization gives you access to items that are shared within your organization. Depending on your organization, you may not see anything here, but it's good to familiarize yourself with this location.

Tour Living Atlas

1. On the blue navigation bar, click Living Atlas.

   In the first tutorial, you added a layer from ArcGIS Living Atlas to your map. It's handy to be able to browse for that content in Map Viewer and add it to your map, but this method is another way to browse and filter content.

2. Browse the ArcGIS Living Atlas content.

3. When you have finished, close your browser.

Summary

In this chapter, you used Map Viewer to make a simple web map. You also explored how content is managed, joined a group, and toured the ArcGIS Online navigation bars.

Workflow

1. Open Map Viewer.
2. On the Contents toolbar, in the Layers pane, search for and add layers.
3. On the Settings toolbar, in the Styles pane, modify the look of the layers.
4. On the Contents toolbar, save and share your map.

CHAPTER 2
Using tables and smart mapping

Objectives

- Explore tables.
- Use different fields to change the subject of your map.
- Try some smart mapping styles.

Introduction

In this chapter, you'll take a closer look at how data is stored and how you can visualize that data with smart mapping.

In the first tutorial, you'll explore how data is stored in a table and how a table is used to make a layer. The second tutorial shows how to choose fields from the table to style your map.

Smart mapping evaluates the data you want to display and suggests styles that may work best. Choosing from these style options helps you quickly create beautiful and informative maps. With smart mapping, you have access to color options designed by color theory experts, map symbols for any scale, and styles that are appropriate for your data.

Tutorial 2-1: Use tables to work with layers

Layers can be thought of as a graphical display of data, which is stored in a table. These tables are made up of rows and columns, where each row represents a feature that will appear in the layer and each column represents an attribute.

You'll view the item details of a layer, review a table, and see the connection between the layer and the table.

Open a map in Map Viewer and add a layer

1. Open a browser, sign in to **www.arcgis.com**, and open a new map.

> **Hint:** If needed, you can review the detailed steps for signing in and opening a new map at the beginning of chapter 1, tutorial 1-1.

2. In the Layers pane, click Add.

> *Review the basics for navigating the user interface in Map Viewer. The Contents toolbar is the dark vertical toolbar on the left, and the Settings toolbar is the light vertical toolbar on the right. When you click or select something on a toolbar, a pane usually opens, allowing you to adjust a related setting. Sometimes, clicking something in a pane opens an additional pane that shows more settings.*

3. In the Add layer pane, click My content, and change the selection to ArcGIS Online.

   Instead of adding content you own (from My content), you will search for public content in ArcGIS Online.

4. Near the top of the pane, in the search box, type **owner:Top20EssentialSkillsForArcGISOnline Colorado**.

> **Hint:** The first part of this search term, "owner:Top20EssentialSkillsForArcGIS-Online," locates all the public content owned by the account that was created for this book. By adding "Colorado," you narrow the search to find a specific item with that term. Throughout this book, you'll use this search method to find items.

5. Click the Colorado Housing Cost layer name to review the layer before adding it.

Before you add a layer, you can see a brief overview of it. This layer provides key details about housing in census tracts around Denver, Colorado. The data is taken from the American Community Survey, a yearly snapshot of demographic information. More details about this layer can be seen on the item details page, including the data table.

6. At the bottom of the pane, click View details.

   The Colorado Housing Cost item page opens on a new browser tab.

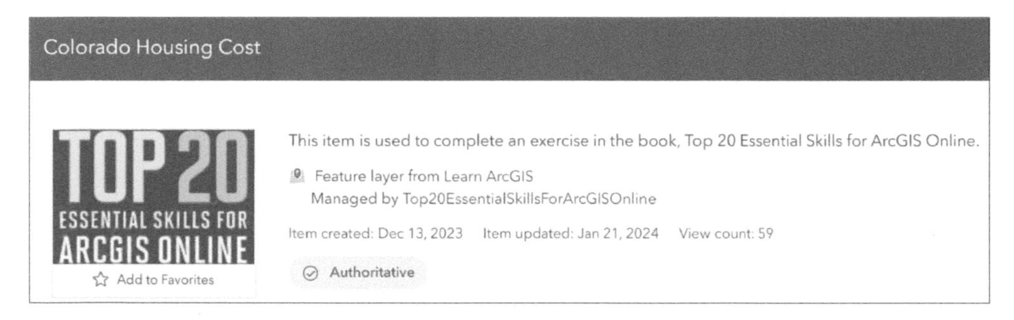

View a table

This page shows the full item details of this layer. Every item in ArcGIS Online has a similar page.

1. Take a second to scan the page and get a sense of the information available.
2. On the blue navigation bar, click Data.

   Now you can see the full table of data that is the basis of this layer. Each row is a census tract feature. Each column lists an attribute about that tract.

3. In the table, hover over the columns of attributes to see what is recorded.

After you add this layer to your map, you can map any of these attributes.

4. Return to the browser tab with your map.
5. At the bottom of the details pane for the Colorado Housing Cost layer, click Add to map and close the pane.

   The layer is added to the map.

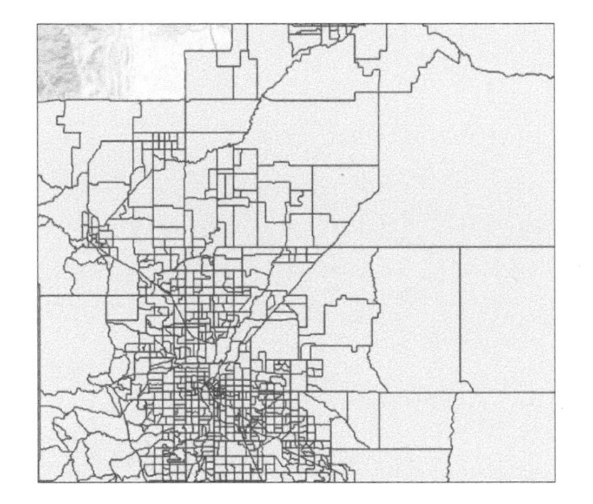

View the table in Map Viewer

1. At the top of the Add layer pane, click the back arrow to return to the Layers pane.
2. In the Layers pane, to the right of the Colorado Housing Cost layer, click Options and click Show table.

Layers
Colorado Housing Cost
Zoom to
Show properties
Show table

The table that appears at the bottom of the screen should look familiar—it's what you reviewed on the item page and it's what's driving the layer.

> **Hint:** Collapse any open toolbars or panes to see the table better.

3. In the Name field, click the up arrow to sort the list in ascending order.

Colorado Housing Cost
849 records, 0 selected

Geographic Ident...	Name	County
08123000100	Census Tract 1	Weld County
08031000102	Census Tract 1.02	Denver County
08031001000	Census Tract 10	Denver County

4. In the left column of the table, on the first row, click to add a checkmark to the gray box and select the row.

5. At the upper left of the table, click the four arrows pointing inward to zoom to the feature you've selected.

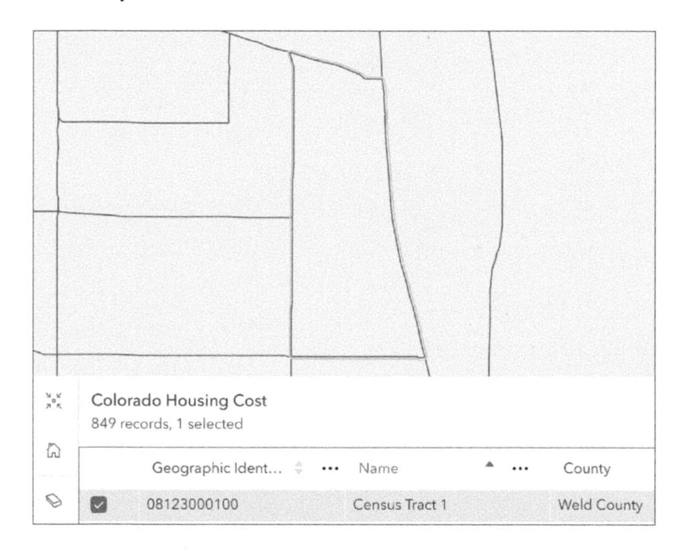

Now you can see the connection between the data in the table and features in the layer. This connection works the other way, as well. You can select a feature in the map and see the row in the table.

6. In the map, zoom out and click the tract just to the south.

7. In the table, on the left, click Show selected.

In the table, the row for that feature is now selected.

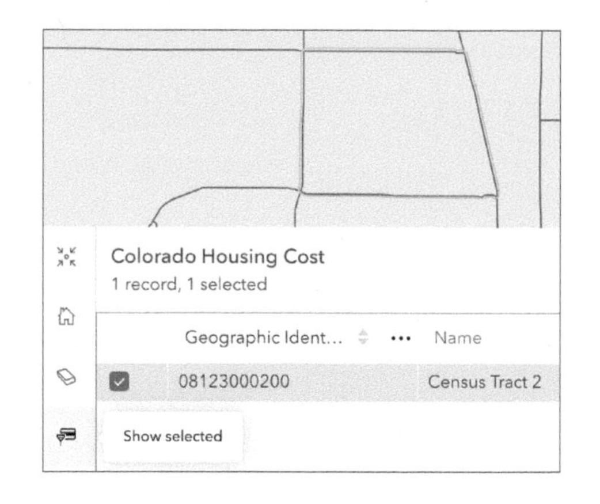

A pop-up appears in the map when you click the tract. By default, the data displayed in the pop-up is the same as what you see in the table for that row. It's easier to see the fields and values in a pop-up rather than scrolling through the table. In a later chapter, you'll learn the essential skill of improving pop-up design.

8. Close the pop-up.

Organize the table

Tables often have many fields, including some that you may not be interested in, so you may want to limit the fields that are displayed. You'll take a closer look at the table and decide what's important.

1. At the upper right of the table, click the up arrow.

2. In the table, on the left, click Show all.

   This expands the table and removes the selection, showing all the rows. If you don't need to see the map at the moment, and you're interested in the table, this is a handy way to organize your workspace.

3. Imagine that you're trying to make a map to show rental units. Review the fields in the table. Consider whether any don't seem relevant to making a simple map. Hover over the fields to see the full description.

   You probably only need to see the Total Renter-Occupied Housing Units (Renter Households) field.

4. At the upper right of the table, click the Settings button (gear icon).

In the Field Visibility pane, notice the blue checkmarks next to the field names. Next, you'll uncheck the fields you don't need, hiding them from view.

5. Uncheck the check boxes for all the fields except Total Renter-Occupied Housing Units (Renter Households). Click Done.

   The table is now focused on only the relevant data. Unchecking the fields doesn't permanently remove anything from your table—it temporarily hides those fields.

6. Review the table and close it.

Tutorial 2-2: Style a map using smart mapping

Smart mapping helps you quickly create effective maps. It analyzes your data table to highlight what's most important in your data. In tutorial 1-1, you learned how the table stored and organized your data. In this tutorial, you'll learn how to choose a field from the table; the field will determine how the layer is styled. Once you've selected one or more fields that you're interested in, smart mapping analyzes the field type and values and provides some great style options to work with.

If you didn't complete the previous tutorial, sign in to your ArcGIS account and open a new web map. In the Layers pane, click Add, click My content, and change the selection to ArcGIS Online. In the search box, type **owner:Top20EssentialSkillsForArcGISOnline Colorado** and add the layer.

Style using Types (unique symbols)

1. On the Settings toolbar, click Styles.

   The Styles pane guides you through the workflow for applying styles. Under step 1 (Choose attributes), no attributes are selected. If no field is selected, a single symbol will show each feature with the same color and outline, which is the default that you're seeing on the map. In the Styles pane, under Pick a style, this style is known as Location (single symbol).

   In tutorial 2-1, the Colorado Housing Cost table had a number of interesting fields, so now you will start to use them.

2. Under Choose attributes, click Field.

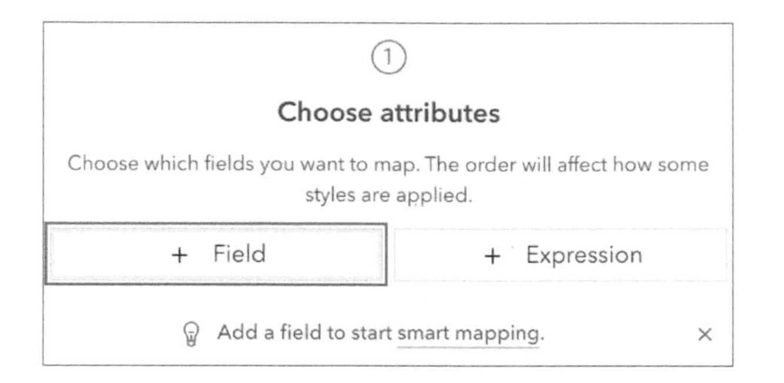

3. In the Select fields list, next to the County field, click the small *i* for additional information.

   The Field type is String, and there are 849 records in this layer, with no empty records.

   Every feature *should* have the County recorded. Empty records may indicate missing values, so it's good to check this and confirm that the data is dependable.

4. Click the back arrow to return to the Select fields pane.
5. Select the County field and click Add.
6. Zoom out on the map.

   Because you have selected a single attribute, smart mapping has suggested the style Types (unique symbols). Each county is now symbolized in a different color.

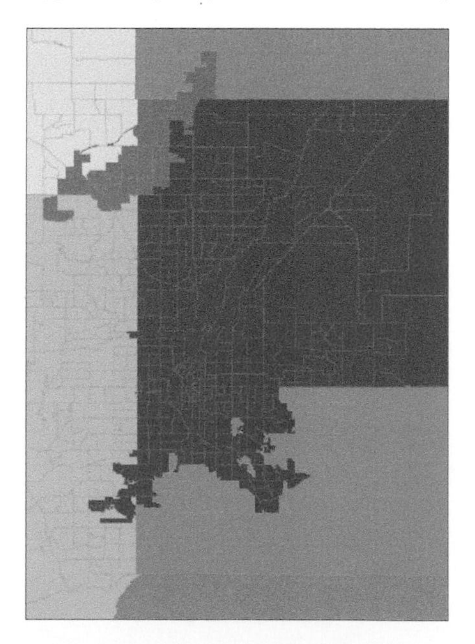

> *Each style can be refined in the Styles pane. Different cartographic methods should be used depending on the data you are displaying. You will learn about that later. For now, you will continue exploring other styles.*

7. Zoom and pan to explore the map.

Style using Counts and Amounts (size)

1. In the Styles pane, under Choose attributes, click County.

   The Replace field pane opens, letting you choose a different field.

2. In the Replace field pane, next to the Total Renter-Occupied Housing Units (Renter Households) field, click the small *i* to review the field type.

   The field type is Integer. This will determine the style options that you'll see later.

3. Click the back arrow to return to the Replace field pane.

4. Click the Total Renter-Occupied Housing Units field.

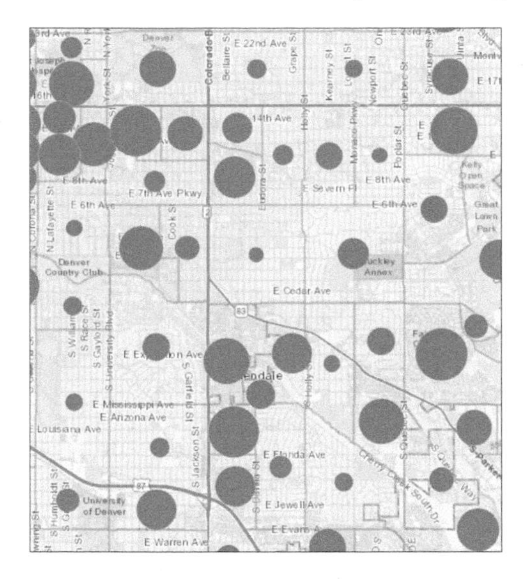

   Smart mapping recognized that the field type was integer and automatically selected a Counts and Amounts (size) style.

   You'll see one circle for each census tract. The bigger the circle, the more rental units are in each tract. The circles may be overlapping a bit, but you can see that certain tracts have high numbers of renters.

5. Zoom and pan to explore the map.

Style using Dot Density

1. Under Pick a style, click Dot Density.

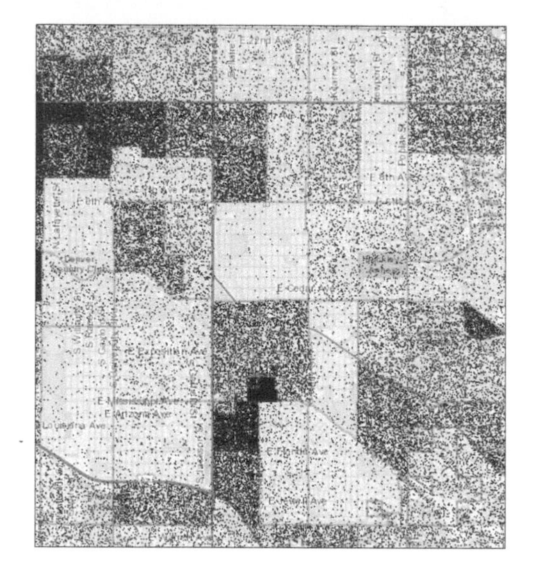

Depending on the scale you are viewing the map at, one or more randomly placed dots are being used to represent the number of rentals in each tract. This visualization helps you clearly see areas of high density.

2. Zoom and pan to explore the map.

Style using Counts and Amounts (color)

You'll use a field showing a percentage and view a suitable style for that type of data.

1. In the Styles pane, under Choose attributes, click Total Renter-Occupied Housing Units (Renter Households).

   The Replace field pane opens, letting you choose a different field.

2. In the Replace field pane, click Percent of Renter Households for whom Gross Rent (Contract Rent Plus Tenant-Paid Utilities) is 30.0 Percent or More of Household Income.

   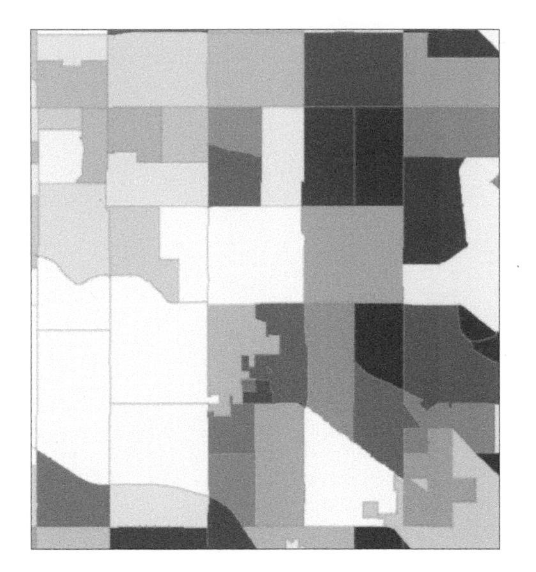

   In the Styles pane, under Pick a style, Counts and Amounts (color) is now selected. This style uses graduated colors to represent the percent of renters who spend a high amount of their income on rent.

3. In the legend, read the percentage range indicated for each tract.

   The values from the field are applied to a color range. Maps that use this style are referred to as choropleth maps.

Style using size and color combined

You'll use two fields to make a layer with size and color combined.

1. In the Styles pane, click Field and click Total Renter Occupied Housing Units (Renter Households). Click Add.

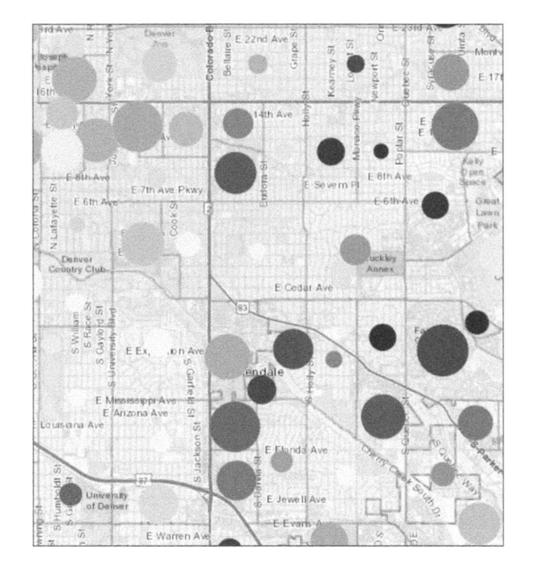

Your layer uses both size and color to communicate a message. Circle size indicates more or fewer people living in rental housing. Color indicates whether many or few renters spend a lot on rent. For example, big, colorful circles indicate areas with a lot of renters who spend a lot on rent.

2. Zoom and pan to explore the map. See if you can find any interesting patterns, for example areas with high numbers of renters who don't seem to spend too much of their income on rent.

Save and share your map

1. On the Contents toolbar, click Save and open, and click Save as.

2. In the Save map window, apply the following settings:

   - Title: **Smart Mapping <your initials>**
   - Folder: Top 20 Tutorial Content
   - Tags: **Top 20, Smart mapping**
   - Summary: **A map showing Smart Mapping styles**

> *If you need to create the folder, click the down arrow and choose Create new folder. Type* **Top 20 Tutorial Content**.

3. Click Save.
4. On the Contents toolbar, click Share map.
5. In the Share pane, click Everyone (public) and click Save.

Take the next step

This section is your chance to try something on your own and explore a bit. There are no step-by-step instructions, just a suggestion to try something fun.

Look over some of the other fields in the layer, and pick a style.

In a later chapter, you'll learn more about cartography, but for now be creative and explore the style options. In the Styles pane, under the applied style, click Style options. Change colors or modify some of the other settings—have fun!

Summary

In this chapter, you learned how tables are used to organize data. You saw that layers can be based on any field in a layer. Smart mapping can read the field type and prompt you with plenty of options to choose from.

Workflow

1. Add a layer.
2. Choose fields.
3. Pick a style.

CHAPTER 3

Creating layers

Objectives

- Create a layer from a KML file.
- Create a layer by sketching inside Map Viewer.
- Turn a spreadsheet into a layer.
- Copy a layer and make it your own.
- Define your own blank layer.
- Configure item settings and share your map.

Introduction

Adding your own layers to a map gives you the chance to add data that's important to you. It's an amazing opportunity to see where something is and share it with the world.

The first tutorial will guide you through a variety of ways to create and add a layer. In the second tutorial, you'll learn the best practices for keeping track of layers and other content you acquire.

> *This chapter covers creating vector layers. Adding raster layers is a somewhat different workflow and is covered later in the book.*

Tutorial 3-1: Create your own layers

In this tutorial, you'll learn how to make your own layers using the methods listed in the chapter objectives. Yep, this tutorial is packed, but depending on where your data is coming from (or whether you're making the data yourself), different methods are needed. The good news is they're all surprisingly simple. With these skills under your belt, you'll be able to make a layer from whatever source comes your way.

Add a layer from a KML file

You may have an app on your phone that you use to track where you go when you're out driving, running, or hiking. Many apps will let you download your path, often as a Keyhole Markup Language (KML) file. Transforming that file into a layer in a web map is an easy way to turn your data into a layer.

You'll download a KML file of a car trip, turn it into a layer, and add it to a map.

1. Sign in to your ArcGIS organization.
2. On the top navigation bar, click the magnifying glass to search for items in ArcGIS Online.
3. Type **owner:Top20EssentialSkillsForArcGISOnline create layers**, and press Enter.
4. On the left, under Filters, turn off the filter for your organization and expand your search to all of ArcGIS Online.
5. On the Create Layers item, click Download to download the CreateLayers.zip file to a convenient location on your computer.
6. Unzip the file.

   > *Two files are extracted. You'll use the TravelRoute.kml file now and the PretoriaPoints.csv a bit later. You will upload these files and turn them into layers. You won't need them after that, so don't worry about keeping them around. You can delete them from your computer after you're done.*

7. At the top of the web page, click Map.
8. In the Layers pane, click the down arrow next to Add and choose Add layer from file.
9. In the Add layer window, click Your device, and browse to the TravelRoute.kml file that you extracted from the zip file or drag it into the window.

10. In the window, for Title, type **Travel Route <your initials>**.

> *If you've already created the Top 20 Tutorial Content folder, select it. If you need to create it, click the down arrow and click Create new folder. Type* **Top 20 Tutorial Content**

You would normally supply additional metadata as you save a layer, but for now you can leave them blank. In this chapter's second tutorial, you'll fill out these fields for your new layers all at once.

11. Click Create and click Add to add it to the map.

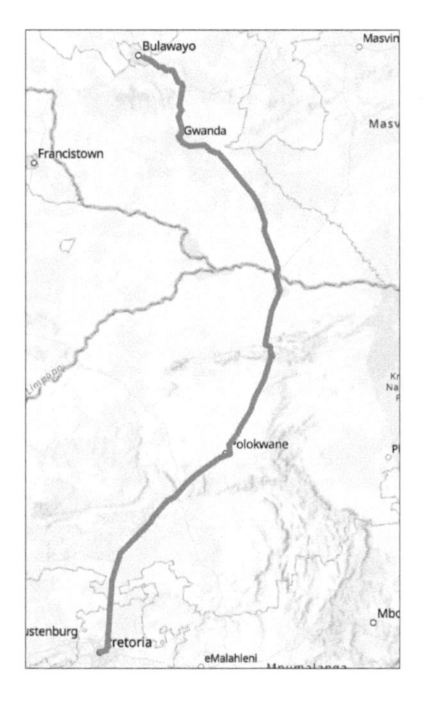

Just like that, the file is turned into a layer. This layer depicts a drive from the National University of Science and Technology in Bulawayo, Zimbabwe, to the National Zoological Gardens in Pretoria, South Africa.

Create a sketch layer

You can add your own layer in Map Viewer by sketching in the map. You can add stamps, lines, polygons, rectangles, circles, and text.

You'll change the basemap and sketch a polygon by tracing the boundaries of the National Zoological Garden in South Africa.

1. On the Contents toolbar, click Basemap. In the Basemap pane, choose Light Gray Canvas.
2. Pan to the southern end of the TravelRoute layer to see the National Zoological Gardens.
3. On the Contents toolbar, click Layers.
4. In the Layers pane, click the down arrow next to Add and choose Create Sketch layer.
5. To the left of the Sketch pane, on the sketch toolbar, choose the Polygon tool (third from the top).
6. On the map, click to begin adding points, and trace the perimeter of the National Zoological Gardens. Double-click to finish the sketch.

> **Hint:** If you hold down the left mouse button, you can continually add points without clicking and sketch more quickly. Press Esc if you need to cancel the sketch and start again.

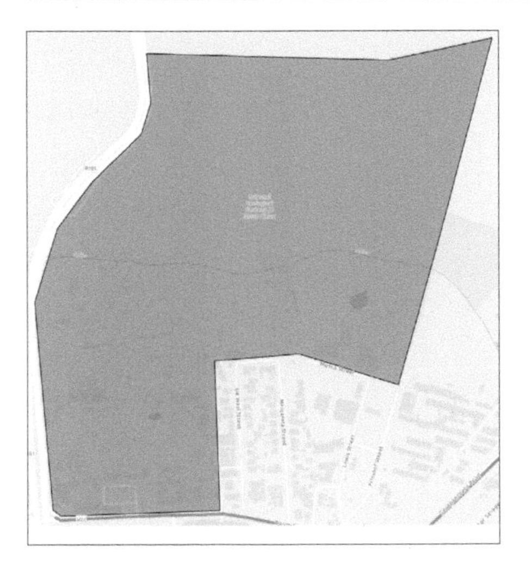

Adding sketches directly to the map is a fast way to add features that you, as the map author, can edit and save.

On your own

Try a quick way to search for a location and turn that search into a point. On the map, click the search tool (magnifying glass). In the search box, type **National Zoological Gardens, Pretoria** and press Enter. In the search results, click Add to make a new sketch. In the Sketch pane, under Current symbol, click Vector point, and from the Animals category, choose a symbol. Click Done.

Create a layer from a spreadsheet

Spreadsheets are useful for organizing rows of valuable information, but often they would be easier to understand if the data were displayed on a map.

You'll use a spreadsheet of points of interest with columns for latitude and longitude and add it to the map as a layer.

1. In the Layers pane, click the down arrow next to Add, and click Add layer from file.
2. In the Add layer window, click Your device and browse to the PretoriaPoints.csv file that you extracted from the zip file, or drag it into the window.

   > *The steps for locating, downloading, and extracting the PretoriaPoints.csv file are at the beginning of this chapter.*

3. Confirm the option to create a hosted feature layer and add it to the map. Click Next.
4. Confirm that the three fields in the CSV file that will be included in the layer have been automatically identified. Click Next.
5. Confirm that the latitude and longitude fields that will be used to place the points in the map have been correctly identified. Click Next.
6. In the window, add the following details:
   - Title: **Pretoria Points of Interest <your initials>**
   - Folder: Top 20 Tutorial Content
7. Click Create and click Add to add the layer to the map.

8. Pan around the map using your mouse. Click some of the five points of interest that have been transformed from a spreadsheet into a layer.

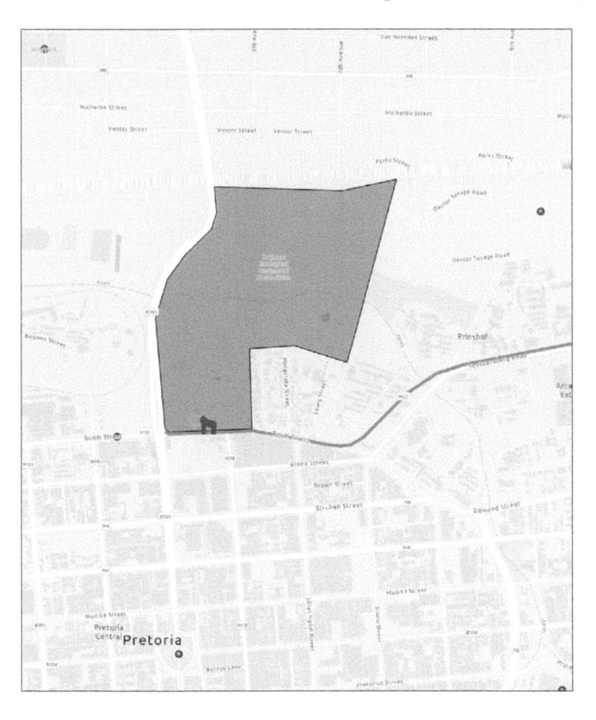

Copy a shared layer

Next, you'll locate a public layer that has been shared by the book's ArcGIS Online account and save a copy.

1. Zoom and pan so that the National Zoological Gardens fills your view.
2. In the Layers pane, click Add.
3. Click My content and change the selection to ArcGIS Online.
4. In the search field, type **owner:Top20EssentialSkillsforArcGISOnline River**.
5. On the Apies River item in the search results, click Add.
6. At the top of the pane, click the back arrow to return to the Layers pane.
7. In the Layers pane, to the right of the Apies River layer name, click Options and click Save as.

8. In the window, add the following details:
   - Title: **Apies River <your initials>**
   - Folder: Top 20 Tutorial Content
9. Click Save.

   The copy is added to your contents and is now used in the map.

> Your Apies River <your initials> layer is a copy that references the Apies River layer shared by the book account. If the layer from the book account were to be deleted or no longer shared, your copy of the layer would no longer be available.
>
> Additionally, you may set properties of your layer to be stored in the map or the layer, depending on your preference. For example, changes to the style can be saved to the layer so that if your layer is added to another map, the layer's style changes would be present in the new map. If you store style properties in the map, changes to the style appear only in that map. You may also discard your changes to the properties and set your layer to use the original properties of the layer that's being referenced. In this scenario, if the style of the layer shared by the book account were updated, the changes would appear in your layer, as well.
>
> For more information about these settings in ArcGIS Online documentation, go to links.esri.com/ArcGISOnlineDocumentation.

Save the map

Now that you've added several layers to your map, take a moment to save it.

1. On the Contents toolbar, click Save and open, and click Save as.
2. In the Save map pane, add the following details:
   - Title: **National Zoological Gardens <your initials>**
   - Folder: Top 20 Tutorial Content
3. Click Save.

   Next, you'll view your Content.

4. In Map Viewer, at the upper left, click the menu button (three horizontal lines), and click Content.

Define and create a blank layer

So far, you've learned a few methods for creating layers in Map Viewer. This time, you'll work from your Contents page. You'll create a hosted feature layer with fields that you define. Then you'll open this layer in Map Viewer and add features for roads in Pretoria.

1. In the upper left on the Content page, click New item.
2. In the New item window, click Feature layer.
3. Under Select an option to create an empty feature layer, confirm that Define your own layer is selected, and click Next.

> **Hint:** For a future project, choosing a template at this step may help save some time. Templates provide preconfigured layers, designed for specific use cases.

4. Under Specify name and type, delete the existing text and type **Roads <your initials>**. Click Point layer, and change the layer type to Line layer.

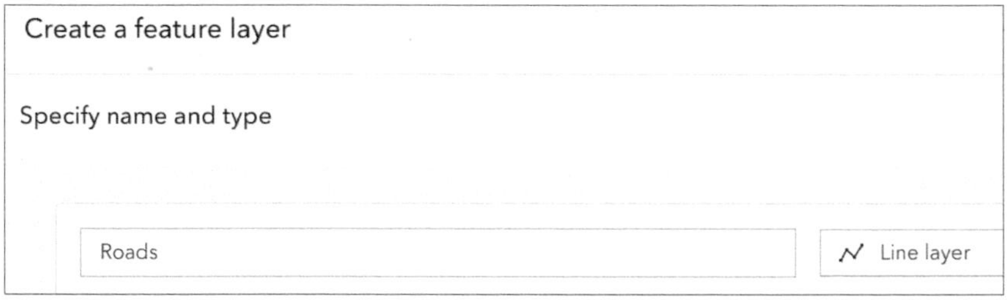

5. Click Next.
6. Confirm the Title. For Folder, click the down arrow and choose Top 20 Tutorial Content. Click Save.

A new hosted feature layer is created and added to Contents. The item page is displayed.

Add a field to your new layer

Now that you have published a hosted feature layer, you'll add a field to store information about the features being added.

1. On the Roads <your initials> item page, on the blue navigation bar, click the Data tab.

   The Data tab lets you add fields and build a table.

2. In the upper right, click the Fields tab.

   Three default fields—*OBJECTID*, *Shape_Length*, and *Photos And Files*—are shown. You'll add a Name field, which will let you store the name of each road feature being added.

3. In the upper left, click Add.
4. Fill out the Add Field window as follows:
   - Field Name: Roads
   - Display Name: Roads
   - Type: String
   - Length: 256
5. Click Add New Field.

   The Roads field is added to the list of fields.

Reopen the National Zoological Gardens map

Now that you've created a new layer, you'll reopen the map you were working on so you can add the Roads layer.

1. On the top navigation toolbar, click Map.

   A new, blank map appears.

2. On the Contents pane, click Save and open, and click Open map.
3. In the list, find the National Zoological Gardens map, and click Open map.

Add the Roads layer

1. In the Layers pane, click Add.
2. In the Add Layers pane, find the Roads <your initials> layer and click Add.
3. On the Settings toolbar, click Edit.

   Now you can edit existing features or add new ones.

4. In the Editor pane, under Roads <your initials>, click New Feature.
5. In the map, zoom to the south of the National Zoological Gardens and find Margareta Street.
6. Click to start a line feature and trace Margareta Street. Double-click to finish adding the feature.
7. In the Create features pane, under Roads, type **Margareta St.** and click Create.

   Roads

   Margareta St.

This is the basic workflow for creating a hosted feature layer, adding a field, and making new features.

Explore the final map

The final map shows different layers that were created using various methods.

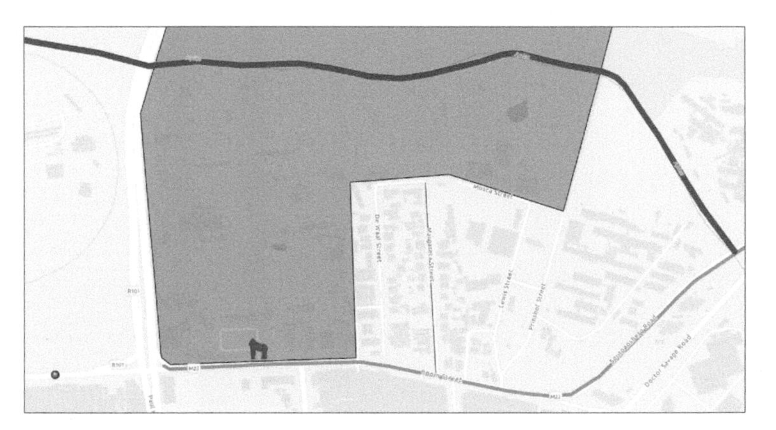

1. Zoom and pan to explore the layers.

On your own

In the Styles pane, try out a few style options.

2. On the Contents toolbar, click Save and Open > Save.

Now you can add layers to a map using various methods, depending on the data you have available.

Tutorial 3-2: Configure item details and settings

In tutorial 3-1, you created several items. Before you saved each item, you were prompted to enter information such as Title, Location, and Summary. Now is your chance to fill out the item details so you'll know what each item is and why it was made. Adding details as soon as you create a new item may be the best way to keep track of the great content you make.

Complete the item details

You'll open the Top 20 Tutorial Content folder containing items you created for this book. You'll open the items that you made in this chapter in new browser tabs, complete any missing item details, and share the items.

1. In Map Viewer, in the upper left, click the menu button and click Content.
2. On the left, under Folders, click Top 20 Tutorial Content.

   Locate the items you created for this chapter.

   You can see the type of item, a symbol indicating whether the item has been shared, an options button, and the date that the item was last modified.

3. Find the web map you made (National Zoological Gardens), right-click View details, and open it in a new browser window.
4. Open the new window and view the map's item details.

   A thumbnail of the layer has been added by default, but all the other fields are blank.

On your own

On the item page for the web map, create a new thumbnail. Click Edit thumbnail, and click Create thumbnail from map. Zoom to a new extent and click Save.

5. On the Item page, at the top, click Add a brief summary about the item, and type **This item was made in chapter 3 of the book Top 20 Essential Skills for ArcGIS Online**. Click Save.
6. For Description, click Edit, and type **This item was created while trying out different ways to create and add layers.** Click Save.
7. For Tags, click Edit, and type **Top 20, ArcGIS Online, Chapter 3**. Separate each tag entry with a comma. Click Save.

On your own

Open browser windows (or tabs) for the remaining items and enter item details. Add a line or two to describe each item in a bit more detail. Save each detail. When you've finished, close the windows and return to the Content/My content page.

Disable editing for the Roads layer

The Roads layer has editing enabled, so you'll turn off editing. You'll need to do this before you can share the layer.

1. On the Roads item, click View details.
2. On the blue navigation bar, click Settings.
3. Under Editing, clear the Enable editing check box.
4. Click Save.

Share your map and layers

Now you're ready to share your map publicly.

1. On the top navigation bar, click Content.
2. In your content list, find the National Zoological Gardens web map and click the button (a person icon) to adjust the sharing level.
3. In the Share window, click Everyone (public) and click Save.

   You're sharing a web map that uses a number of layers that haven't been shared yet, so you're prompted to share all the layers, too. Sharing this way makes it easy to ensure that everything has the necessary sharing settings.

4. Click Update sharing.

   In your content list, notice that the sharing symbols for your map and the layers have been updated to a globe, indicating that they are now public.

Take the next step

Make another layer using one of the methods you learned and add it to a new map. Fill in all the item details for the layer and the map. Share the map.

Summary

In this chapter, you learned how to make layers in several different ways. Getting comfortable with these workflows will give you confidence in your ability to quickly add a layer to a map.

Workflows

Add a layer from a KML file

1. In the Layers pane, click the down arrow next to Add and choose Add layer from file.
2. Add the KML file.

Create a sketch layer

1. In the Layers pane, click the down arrow next to Add and choose Create Sketch layer.
2. To the left of the Sketch pane, from the sketch tools list, choose a tool and create a sketch.

Create a layer from a spreadsheet

1. In the Layers pane, click the down arrow next to Add and choose Add layer from file.
2. In the Add layer pane, add the CSV file.

Copy a shared layer

1. In the Layers pane, to the right of the layer, click Options and choose Save as.
2. Rename the item.

Define and create a blank layer

1. In the upper left of the Content page, click New item.
2. In the New item window, click Feature layer.
3. In the Create a feature layer window, confirm that Define your own layer is selected and click Next.
4. Name the layer and choose the geometry type.
5. On the item page, click the Data tab to see the table and add fields.
6. In the upper right, click the Fields tab to add a field. Fill out the field details.
7. Edit the layer in Map Viewer, adding features and attributes.

CHAPTER 4

Configuring layer pop-ups

Objectives

- Add a feature layer to a new map.
- Configure fields by changing the display name and formatting.
- Use ArcGIS Arcade to format attributes.
- Configure a pop-up title.
- Add pop-up elements, including text, images, and charts, and learn how to create a media group.
- Use HTML to create a table.
- Save your pop-up configuration.

Introduction

Authoring a great web map requires some thought and craft, making use of styles, effects, blending, and other techniques to create a meaningful map. Pop-ups are an equally important part of map authoring, completing the map information experience.

In this chapter, you'll learn how to configure pop-ups using pop-up elements such as text, images, and charts. You'll also learn how to use ArcGIS Arcade in pop-ups for formatting text and calculations. Finally, you'll learn how you can save your pop-up configuration to the layer, making your work and craft reusable in more than one map.

Tutorial 4-1: Configure pop-ups and format pop-up elements

Pop-ups can transform an otherwise dull list of attributes into a far more meaningful display of engaging information for an audience. As a map author, you'll make decisions about pop-ups that reflect the information that's displayed and the best way to present it.

Open a map in Map Viewer and add a layer

You'll add an unconfigured feature layer of mountains in California that are over 14,000 feet high ("fourteeners") and examine the default pop-up.

1. Sign in to **www.arcgis.com** and open a new map.

> **Hint:** If you need, review the detailed steps in chapter 1 for signing in to ArcGIS Online and opening a new map.

2. In the Layers pane, click Add.

> *Review the basics for navigating the user interface in Map Viewer. The Contents toolbar is the dark vertical toolbar on the left, and the Settings toolbar is the light vertical toolbar on the right. When you click or select something on a toolbar, a pane usually opens, allowing you to adjust a related setting. Sometimes, clicking something in a pane opens an additional pane that shows more settings.*

3. In the Add layer pane, click My content, and change the selection to ArcGIS Online.
4. In the search box, type **owner:Top20EssentialSkillsForArcGISOnline Fourteeners**.
5. Next to the California Fourteeners layer, click Add.
6. In the Add layer pane, click the back arrow.
7. On the map, click any feature to view the default pop-up.

   In Map Viewer, pop-ups are enabled on feature layers by default.

Hint: To disable pop-ups in a layer, in the Layers pane, select the layer. On the Settings toolbar, click Pop-ups, and turn off the Enable pop-ups toggle.

The default pop-up appearance is a list of fields and values. The pop-up also includes a title that displays the layer name and first text field (in this case NAME) from the list of attributes.

By default, pop-ups are undocked and are tethered to the feature you click, but pop-ups can be docked to display more information and move them away from the features on the map.

8. In the upper right of the pop-up, click Dock.

Dock

Configure fields

If you plan on using a list of fields and values in your pop-up, it's a good practice to make default field names more readable and understandable by changing the display name. If you're not intending to display the field names in the pop-up, this is optional, but still a good practice, especially if others will use the layer.

Format your field values to properly display the data.

1. On the Settings toolbar, click Fields to display the list of fields.
2. In the Fields pane, click the DESC_ field to edit the Display name. Type **Description**, which is more meaningful and readable.
3. Click Done.

On your own

Change ELEV_FEET to Elevation (feet). Choose new field names as you think best. For numeric fields, you can use the thousands separator to specify the number of significant digits to use. Configure the rest of the field list.

You can select multiple fields of the same type to configure certain formatting at the same time.

Use ArcGIS Arcade expressions to format values

Arcade is an expression language that is supported across ArcGIS. It can be used to perform calculations, manipulate text, and evaluate logical statements.

In pop-ups, Arcade is used in two ways: to create expressions in text elements that can be evaluated like fields, or as separate elements that return a block of content.

1. In the pop-up, notice that the name of the peak is in all capital letters.

   Next, you'll use Arcade expressions to change the text formatting for NAME from all capital letters to something more suitable.

2. On the Settings toolbar, click Pop-ups.
3. In the Pop-ups pane, click Attribute expressions.
4. Click Add expression to open the Arcade editor window.
5. At the top, in the New Expression field, delete the existing text and type **ProperName**.

   This name provides a hint as to what the expression will do, which is to return a properly formatted peak name.

   Next, you'll use the Proper function to change the case for the NAME field. The name (NAME) for every feature can be accessed using `$feature`.

6. Delete the example text in the expression window and type the following expression:

```
// Returns proper name for the peak
// Change all caps to proper text
var txtProper = Proper($feature.NAME, 'everyword')
return txtProper
```

> *The larger blocks of code used in this chapter are provided in a document that you can use to copy and paste from. You can download the document at links.esri.com/Top20Code.*

ProperName

▷ Run

```
1  // Returns proper name for the peak
2  // Change all caps to proper text
3  var txtProper = Proper($feature.NAME, 'everyword')
4  return txtProper
```

This expression will return the proper case text for NAME. If NAME contains the string `MOUNT WHITNEY`, the expression returns `Mount Whitney`.

7. Click Run to verify the expression, and click Done.
8. In the Attribute expressions pane, click Back.

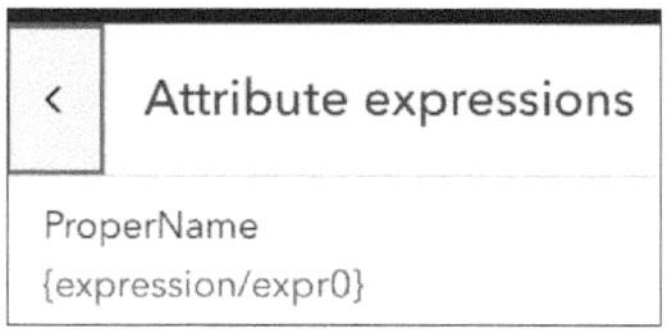

< Attribute expressions
ProperName
{expression/expr0}

Configure the pop-up title

The default pop-up title is a combination of the layer name and the first text field—in this case NAME. In this section, you'll edit the pop-up title to use the expression created earlier to return the proper text string.

1. In the Pop-ups pane, click Title to begin editing.

   Fields and expressions are displayed inside curly braces. The field value shown in the Title field will be shown for each feature's pop-up. For example, if you click the Mount Whitney feature in the map and the pop-up Title field has the expression `{NAME}`, the pop-up will display `MOUNT WHITNEY` in the title.

   You'll change the title field to use the Arcade expression you created, displaying the proper text-formatted name instead of the uppercase field value.

2. Delete the existing text and click the curly braces to the right of the title input field.
3. In the Add field pane, click ProperName.

   When you select an expression, it's displayed using its internal name `{expression/expr0}`.

4. Preceding the expression, type **California Fourteeners:** with a colon, as shown.

   Your final title should look like the following figure.

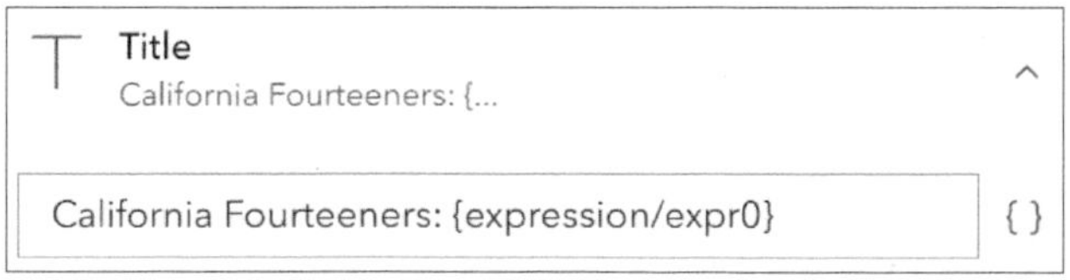

Title
California Fourteeners: {...
California Fourteeners: {expression/expr0} { }

Create a text element

A pop-up consists of one or more elements: Fields list, Chart, Image, Text, and Arcade. Text can be configured using one or more fields or expressions and can also include free-form text that you can configure using different sizes, fonts, and colors.

You'll add a text element using a combination of free-form text and fields.

1. In the Pop-ups pane, click the More button (ellipsis) next to Fields list and click Delete.

   The current pop-up content is deleted, allowing you to design your own.

2. Click Add content and click Text.

   The text editor appears. You'll add a combination of free-form text and fields.

3. Type an open curly brace to display the list of fields. From the list, click `{ProperName}`.

   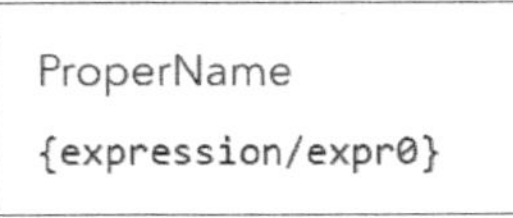

4. Highlight the `{expression/expr0}` field. On the toolbar above the text box, click Font color. Choose a color from the Font color picker.

   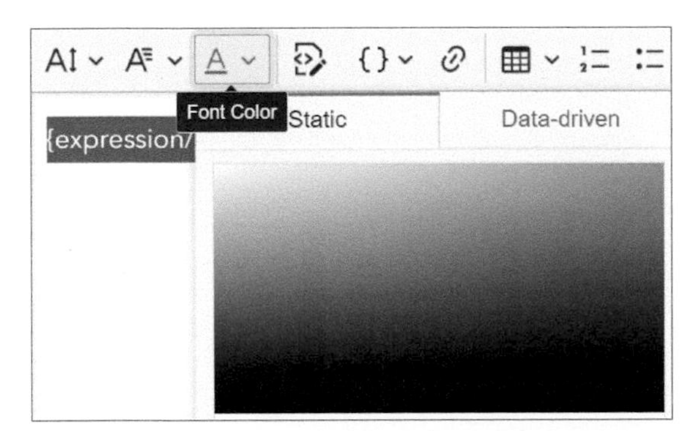

5. Click Done.

On your own

Apply other formatting options such as bold, italic, and so on.

6. Continue typing to create the following expression:

```
{expression/expr0} is a peak in California's {RANGE} range. It
ranks number {RANK} among the California Fourteeners.
The summit is {ELEV_FEET} feet high ({ELEV_METERS} meters) and has
a prominence of {PROM_FEET} feet ({PROM_METERS} meters).
Description: {DESC_}
```

   You'll add the text string `More information` and a link using the `{REF}` field.

7. On a new line, type **More Information** in the text editor and select it.

8. Click Link at the top of the editor.

9. In the link input, type **{REF}**.

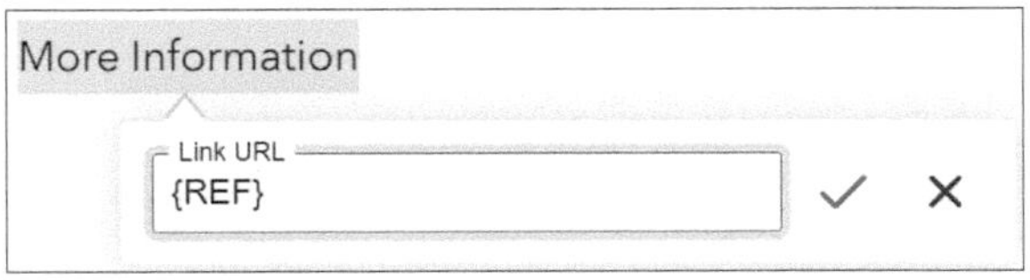

10. Click the green checkmark.

> **Hint:** You can use the curly braces to browse fields, add one to the text editor, copy and delete it, and paste it into the link.

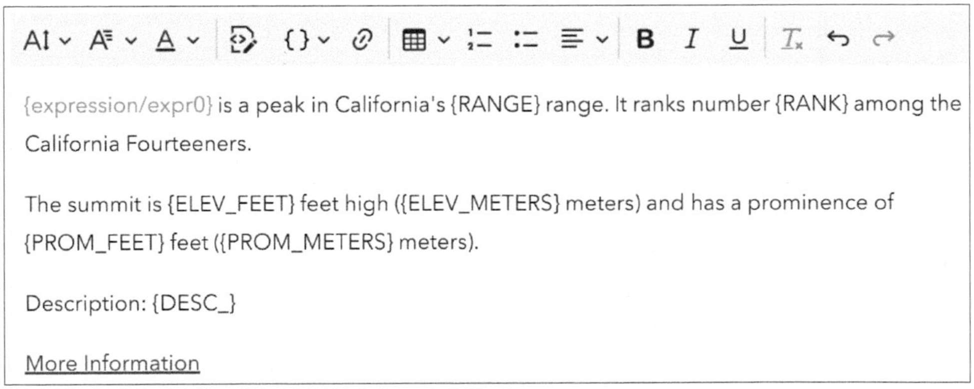

11. Click OK.

Add an image element

Images in pop-ups can add spice and context to a text-only display. In this next section, you'll add an image, including credits for the photo and a link to view a larger version of the image in a new browser window or tab. You can stack images vertically (by default) or combine multiple media elements to create a media group, which is navigated horizontally in the pop-up.

Images can be stored in any online repository, such as ArcGIS Online, Box, Flickr, or others. The image should be publicly accessible, and the URL to the image shouldn't open a viewer app but should instead point to the source image.

Viewing your pop-up images

You can view these publicly shared images in various online repositories:

- ArcGIS Online—links.esri.com/popupArcGISOnline
- Flickr—links.esri.com/popupFlickr
- On a server—links.esri.com/popupServer

> *When using images in pop-ups, ensure that they're sized appropriately for web display. Large images can take a long time to appear in the pop-ups, visibly painting into view from top to bottom. To avoid this and for best performance, optimize the file size for web display. If necessary, create a smaller version of the image for the pop-up display, keeping a larger version as an image that can be opened with a link.*

You'll add a reduced-size image to the pop-up and link to a full-size version.

1. In the Pop-ups pane, click Add content and click Image.
2. In the Configure image pane, next to the URL input, click the curly braces.
3. From the Fields list, click PHOTO_SMALL_URL.
4. In the Title box, type **Click to view larger image**.

   You'll use the Caption box to create a link to the photo source, which will provide credits for the photo.

5. In the Caption box, type **`<a href="{PHOTO_CREDIT}" target="_blank"> Photo credit</a>`**.

6. In the Alternative text box, click the curly braces and click the NAME field.

7. In the Link box, click the curly braces and click the PHOTO_URL field.

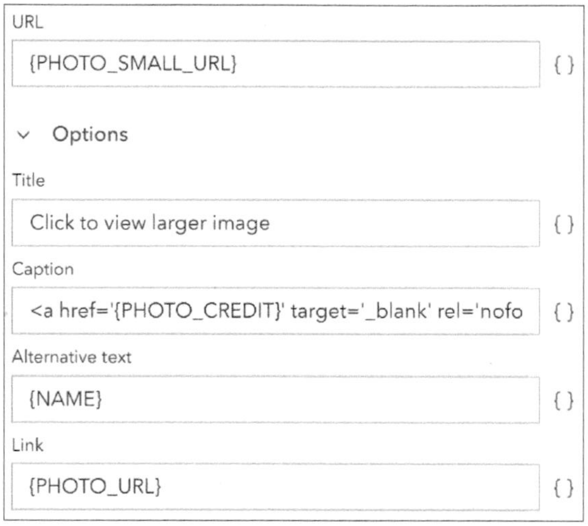

URL
{PHOTO_SMALL_URL} { }
Options
Title
Click to view larger image { }
Caption
<a href='{PHOTO_CREDIT}' target='_blank' rel='nofo { }
Alternative text
{NAME} { }
Link
{PHOTO_URL} { }

8. Click Done.

Add a chart element

Charts in pop-ups provide a meaningful way to display numeric attribute information. You can add bar charts, line charts, pie charts, and column charts.

Bar charts are best used to show data in discrete categories. Spaces between the elements separate the values. Bar charts can be oriented vertically or horizontally.

Line charts can be used to show change over time or a progression. A line chart indicates an inherent order, progressing from left to right.

Pie charts are effective for showing the parts of a whole. All the attributes shown in the pie should add up to 100 percent.

Chart elements, such as image elements, are a type of media element. They can be arranged vertically or arranged horizontally when combined in a media group. You'll add a bar chart showing the relationship between the height of the peak and its prominence, a measure of the elevation of a summit compared with its surrounding terrain.

1. In the Pop-up pane, click Add content and choose Chart.

2. In the Configure chart pane, under Title, type **Elevation vs. Prominence.**

3. Under Alternative text, click the curly braces and select the NAME field.

4. Click Select fields and click ELEV_FEET and PROM_FEET. If you prefer, you can use ELEV_METERS and PROM_METERS. Click Done.

Hint: You can adjust the chart colors. Under Color, click the edit button (pencil) to show the Color Style window. In the Chart style window, choose a new color scheme to match your map or emphasize aspects of the data. You can also adjust the individu al colors by clicking the color dots next to each field in the Configure chart pane.

5. In the Configure chart pane, under Horizontal orientation, click the toggle.
6. In the pop-up, visually confirm the orientation change.

   To better show the difference in elevation vs. prominence, horizontal orientation works best.

7. Choose horizontal orientation, and click Done.
8. On the map, click several features to review the pop-ups.

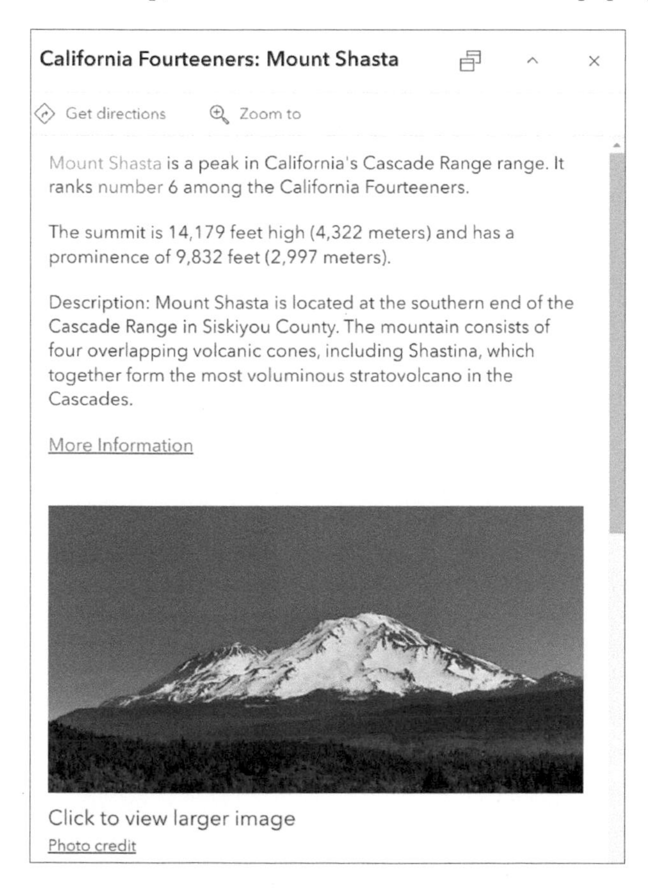

Save your layer

You've now configured a layer pop-up in a web map. If you save this web map, your pop-up configuration will also be saved. In this case, your map will display your pop-up configuration of the California Peaks layer, which is owned by the book's ArcGIS Online account. If others want the same pop-ups in their map, they would need to add the layer from the book's account as you did and recreate the pop-up configuration themselves.

However, if you save a copy of the layer to your account, it will also save your pop-up configuration. You can then share the layer, allowing others to use it and your pop-up configuration. You'll do this next.

1. In the Layers pane, select the California Fourteeners layer.
2. On the Settings toolbar, click Properties.
3. In the Properties pane, expand Information.
4. Click Save as to save the layer to your contents with the pop-up configuration applied.
5. In the Save as window, add the following settings:
    - Title: **California Fourteeners <your initials>**
    - Folder: Top 20 Tutorial Content
    - Tags: **Top 20, ArcGIS Online, Fourteeners**
    - Summary: **A layer showing the fourteen tallest peaks in California using customized pop-ups.**

    > *If you need to create the folder, click the down arrow and choose Create new folder. Type* **Top 20 Tutorial Content**.

6. Click Save.

Save and share your map

1. On the Contents toolbar, click Save and Open > Save as.
2. In the Save map window, add the following settings:
    - Title: **California Fourteeners Map <your initials>**
    - Folder: Top 20 Tutorial Content

- Tags: **Top 20, ArcGIS Online, Fourteeners**
- Summary: **A map showing the fourteen tallest peaks in California.**

3. Click Save.
4. On the Contents toolbar, click Share map.
5. In the Share pane, click Everyone (public) and click Save.

Take the next step

You can do a lot more with pop-ups if you think creatively and use some basic HTML and Arcade skills. In the next sections, you'll learn how to add videos to your pop-ups, create tables, and use Arcade for performing calculations whose categorized results can be displayed in text elements.

Add a video

Videos can be added in several ways. One way is to add a text or an image element with a link that opens the video in another browser window or tab. There isn't a media element specifically for videos, but you can add videos using the HTML capabilities of text elements if you want the video to play directly in the pop-up.

1. In the Pop-ups pane, click Add content and choose Text.

   The text editor appears.

2. On the toolbar, click the Source tool.
3. Type the following HTML to display a video:

```
<p>
    <video autoplay="" width="100%">
        <source src="{VIDEO}" type="video/mp4">
    </video>
</p>
<p>
    <a href="{VIDEO}" target="_blank" >View larger video</a>
</p>
```

4. Click the Source tool again and click OK.

   In the preceding example, the `{VIDEO}` field contains the full URL to the MP4 file: https://downloads.esri.com/agol/blog/california_fourteeners.mp4.

Since autoplay is used, the video will automatically play in the pop-up. You can substitute the word "controls" for "autoplay" to display the video with player controls: `<video controls="" width="100%">`.

Categorize numeric values using text with Arcade

You may want to look at a range of numeric values and categorize them into descriptive text—for example, cold, warm, and hot or small, medium, and large.

Using Arcade, you can categorize numeric values and return a string for use in a pop-up text element. The following example of Arcade takes the prominence in feet and describes it using Low, Moderate, and High depending on the value.

1. Create a new Arcade expression following the steps you used earlier. Name the expression **ProminenceRank** and add the following code:

```
// ProminenceRank
// Returns a string describing the ranking of the prominence for
the peak.
var prom = $feature.PROM_FEET;
var ranking = When (prom < 1000, "Low", prom >= 1000 && prom <
3000, "Moderate", prom >= 3000, "High", "N/A")
return ranking;
```

2. After adding the Arcade expression, add a new text element to your pop-up to show the prominence ranking, as follows:

   The prominence ranking for {NAME} is {expression/expr1}.

   > *After you create the new Arcade expression, it is listed as* {expression/expr1}.

Add a table

Tables allow you to arrange data into rows and columns. The text element editor includes an Insert Table tool that can be used to lay out the table. You can also use the source code editor to add colors and styling to the table. Try adding a table now.

1. Add a text element.
2. In the text editor, click the Insert Table tool on the toolbar.

   A matrix appears, indicating the number of rows and columns the table will have.

3. Drag to create four rows and two columns (4 × 2) needed for the table.

4. Select each cell and enter the following text and field pairs. You can select fields by entering a curly brace.

Name	**{expression/expr0}**
Rank	**{RANK}**
Elevation	**{ELEV_FEET} feet**
Range	**{RANGE}**

> **Hint:** Advance to the next cell using the Tab key

5. Click the first cell (upper left). From the Table tools menu, click Row, and click Insert row above.
6. Click inside the new row. From the Table tools menu, click Row, and turn on the Header row toggle.
7. For Text, type **California Fourteeners**. Select the text and use the Font Size tool to increase the size to **18 pts**.
8. Select the cell with the text. From the Table tools menu, click the Cell Properties tool. Change the background color to light blue and the table cell text alignment to center.
9. Click the cell to the right of the colored cell, click the down arrow next to Merge cells, and click Merge cell left.
10. Click OK.

> **Hint:** You can use the Source tool to examine the HTML created earlier and add additional HTML to further style the table. Alternatively, you can search for online tools and examples to obtain the HTML, paste the source HTML, and modify it with field names.

Summary

In this chapter, you learned how to configure fields and use ArcGIS Arcade to format field attributes. You also learned how to configure pop-up titles and add text elements, images, and charts.

You now have the essential skills to craft pop-ups that turn a list of attributes into meaningful and expressive information that complements your map.

Workflow

1. Open a map in Map Viewer and add a layer.
2. Configure fields.
3. Use ArcGIS Arcade expressions to format values
4. Configure pop-up titles.
5. Create text, image, and chart elements in a pop-up.
6. Save and share layers and maps.

CHAPTER 5
Applying cartography to a web map

Objectives

- Rename and organize layers.
- Style and blend layers.
- Style features based on field values.
- Duplicate a layer.
- Filter features based on an expression.
- Set visible ranges.
- Create labels.
- Add an effect.
- Edit a legend.

Introduction

In this chapter, you'll learn basic cartographic techniques for making a web map. You'll organize, style, and configure your layers, creating a map that's attractive and easy to use.

Tutorial 5-1: Use cartographic techniques to enhance a web map

First, you'll open a web map of Ireland with all the layers you'll need. Next, you'll organize the layers, ordering them so that they draw logically. You'll style them to communicate clearly, choosing the scale that each layer will be visible at. You'll also apply effects to create an eye-catching map.

Open a map in Map Viewer and add a layer

1. Sign in to ArcGIS Online.
2. On the toolbar at the top, click the magnifying glass to search for items in ArcGIS Online.
3. In the search box, type **owner:Top20EssentialSkillsForArcGISOnline Cartography**. Press Enter.
4. Under Filters, turn off the filter for your organization and expand your search to all of ArcGIS Online.
5. On the Ireland Cartography web map, click Open in Map Viewer.

   The unconfigured map of Ireland opens in Map Viewer.

Explore the web map and rename the layers

You'll rename the layers to make them clear to other users. The top layer, Ireland, is clear, but the others should be renamed to follow a similar naming pattern.

1. On the Contents toolbar, click Layers.

   > *Review the basics for navigating the user interface in Map Viewer. The Contents toolbar is the dark vertical toolbar on the left, and the Settings toolbar is the light vertical toolbar on the right. When you click or select something on a toolbar, a pane usually opens, allowing you to adjust a related setting.*

2. In the Layers pane, at the end of the NorthernIreland layer, click Options > Rename. Edit the layer name to read **Northern Ireland**.

3. Rename the other layers as follows:

    - IrelandCounties: **Counties**
    - Main Roads: **Roads**
    - Ireland_Cities: **Cities**

Reorder the layers

To build the map you're seeing, Map Viewer first draws the bottom layer that's listed in the Layers pane, then draws the second from the bottom and so on, until the layer at the top of the list is drawn. In this case, since the top layer, Ireland, is drawn last, it covers all the layers below it. The result is that only the Ireland and the Northern Ireland layers are visible.

To show all the layers in your map, you'll order them with polygon layers at the bottom of the Layers list, so that they draw first. Line layers will be next, so that lines are drawn on top of the polygons. Finally, point features will be at the top of the list, so that they draw on top of everything.

1. Click the Ireland layer and drag it to the bottom of the list.

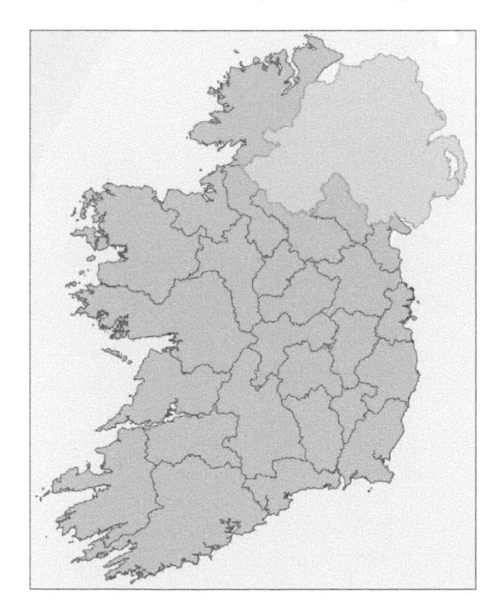

Since the Ireland layer draws first and the Counties layer draws after it, you can now see the Counties layer. The Northern Ireland layer is the only layer in its extent, so there are no underlying layers to consider. Having it draw first doesn't affect any other layers.

2. Continue to reorder the layers from top to bottom, as follows:

    - Northern Ireland
    - Cites
    - Roads
    - Counties
    - Ireland

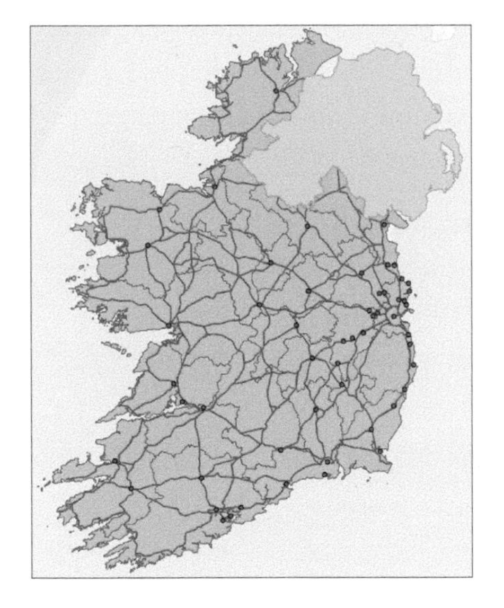

3. Starting at the top of the list, point to each layer and click the eye to turn off layer visibility. Watch the changes in the map as you turn off each layer.

   As you turn off layer visibility, you can see how each layer is drawn on top of the one below it. The basemap is visible once all the layers have been turned off.

Style the Ireland and Northern Ireland layers

The layers have default symbology applied. You'll make them visible and style them, one at a time.

1. Select the Ireland layer and turn on its visibility.
2. On the Settings toolbar, click Styles.
3. In the Styles pane, under Location (single symbol), click Style options.
4. Under Symbol style, click the default.

5. In the Symbol style pane, under Fill color, click the default.

6. In the Select color pane, next to the pound sign (#), enter the hexadecimal value: **cfe6b8**. Click Done.

    Hint: Specific colors can be set quickly by entering a hexadecimal value.

7. In the Symbol style pane, under Outline color, click the default and enter the hexadecimal value: **c8c8c8**. Click Done.

On your own

Select the Northern Ireland layer by clicking it in the Layers pane or clicking the down arrow in the heading of the Styles pane. Change the selection to Northern Ireland. Turn on layer visibility. Style the layer using the Fill color hexadecimal value **dedeab**. For the Outline color, use **c8c8c8**. In the pane, click Done.

The styled layers are shown.

Blend the Ireland and Northern Ireland layers with the basemap

The basemap shows a hillshade, but it's obscured by the Ireland and Northern Ireland layers. You'll blend the layers with the basemap.

1. In the map, zoom to the southwest of Ireland.
2. In the Layers pane, select the Ireland layer.
3. On the Settings toolbar, click Properties.
4. In the Properties pane, under Appearance, adjust the Transparency to **50%**.

   The basemap is now somewhat visible.

5. Under Blending, click the down arrow and click Multiply. Close the pane.

The Ireland layer is blended with the basemap, making the hillshade visible.

On your own

Apply the same transparency and blending settings to the Northern Ireland layer.

Style the Counties layer

1. In the map, zoom out so that all of Ireland and Northern Ireland are visible.

2. In the Layers pane, select the Counties layer and turn on its visibility.

   Counties aren't the main feature of this map, but they can provide helpful context. You'll style the counties so that only the outlines are shown, making them less noticeable and preventing the counties from obscuring the Ireland layer.

3. Open the Styles pane and click Style options.

4. In the pane, click the default.

5. In the Symbol style pane, under Fill color, click No color.

   In the map, only the county outlines are shown, allowing you to see the Ireland layer.

6. Under Outline color, click the default and enter the hexadecimal value **c8c8c8**. Click Done.

7. In the Styles pane, click Done, and review the map.

Style the Roads layer based on field values

The Roads layer has attributes that categorize the roads into different types. Next, you'll group the road types, limiting the number that are displayed, then you'll style the road features with different colors based on these groups.

1. In the Layers pane, select the Roads layer and turn on its visibility.
2. In the Styles pane, click Field.
3. In the Select fields pane, click CWAY TYPE DESCRIPTION. Click Add.
4. To review the table for the Roads layer, on the layer, click Options and click Show table. Find the CWAY TYPE DESCRIPTION field and review the stored values.
5. Close the table.

   The map legend shows the updated symbology, indicating the different types of roads. But it appears that only two road types are used for most of the country.

   2 Lane Road
   Motorway
   Dual Carriageway
   3 Lane Road/1 Lane Side
   3 Lane Road/2 Lane Side
   Type 1 : Dual Carriageway
   Type 2 : Dual Carriageway
   One Way Forward
   One Way Reverse
   Type 3 : Dual Carriageway
   Other

6. In the Styles pane, under Types (unique symbols), click Style options.

   In the pane, each road type is listed with the number of features in it. You'll simplify this list by grouping the road types.

   Only 10 categories can be shown per layer. Otherwise, the map would be difficult to interpret. Categories with the lowest feature counts are automatically grouped in the Other category.

7. Check the Motorway check box, and continue checking boxes for all road types until you reach the Type3 : Dual Carriageway check box. Check that box, too.

8. In the pane, click Move to group.

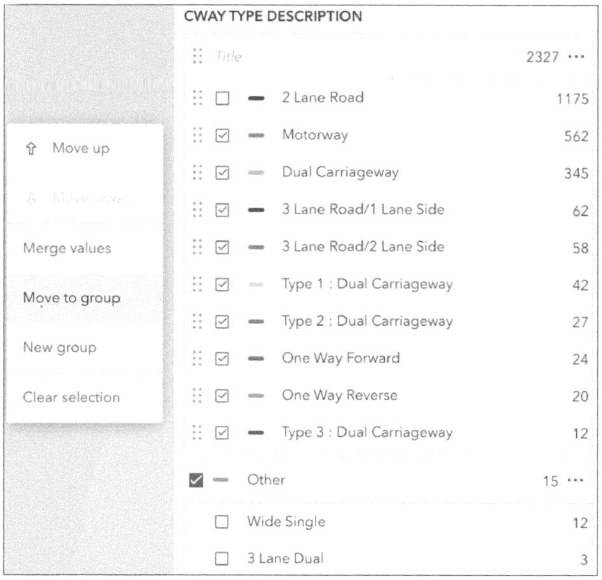

9. In the Move to group window, click Other.

   All road types except the 2 Lane Road type are now grouped in the Other category. You have only two types of road features to symbolize: 2 Lane Roads (the most common) and Other (the remaining types).

10. Next to 2 Lane Road, click the default symbol.

11. In the Symbol style pane, click the default color. In the Select color window, enter the hexadecimal value **545a99**. Click Done.

On your own

Style the Other group symbol using the hexadecimal value **7797d1**. Click Done.

Symbolize the Cities layer

1. In the Layers pane, select the Cities layer and turn on its visibility.
2. In the Styles pane, click Field.
3. In the Select fields pane, click Population, and click Add.

   The Counts and Amounts (size) style is automatically applied.

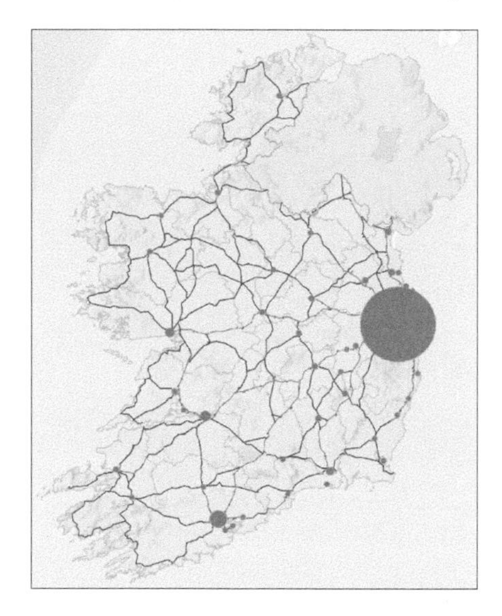

4. Under Counts and Amounts (size), click Style options.
5. In the Style options pane, under Symbol style, click the default symbol.
6. In the Symbol style pane, under Current symbol, click Basic point.
7. Click the Category down arrow. Under Vector Symbols, click Shapes.
8. In the list, click Circle 3, and click Done.
9. In the Symbol style pane, click Fill color and enter the hexadecimal value **a87ee6**. Click Done.
10. Set the Fill transparency to **20%**.

    Adding a bit more transparency will help you see through the city symbol.

On your own

Set the Outline color using the hexadecimal value **594080**, and set the Outline width to **0.5 px**.

11. Close the Symbol style pane.

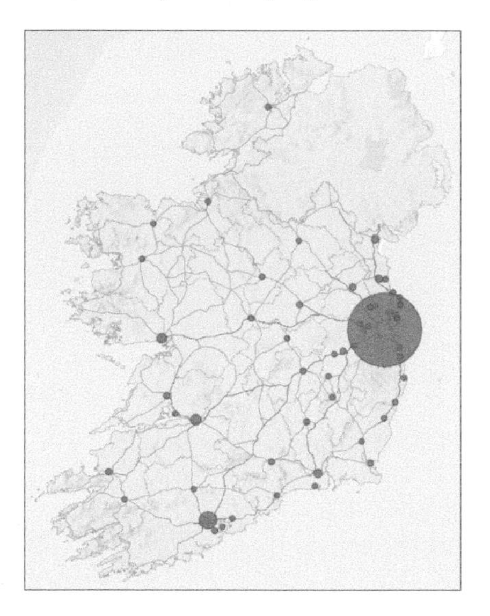

Set the Cities symbol size, based on the Population field

The Cities layer has a Population field. You'll use that field to style the point symbol, visualizing the population size.

1. In the Styles pane, turn on Classify data.

   Classifying your data divides it into groups that you define by setting the ranges and class breaks.

   Dublin is the only city with a population above 500,000. Limerick and Cork are next in size, with populations between 100,000 and 500,000. Remaining cities contain populations under 100,000. You'll set the symbol sizes to reflect that.

2. Under Number of classes, choose 3.
3. In the Data range section, click the lower number on the left and type **100000**.
4. Click the upper number on the left and type **500000**. Press Enter.

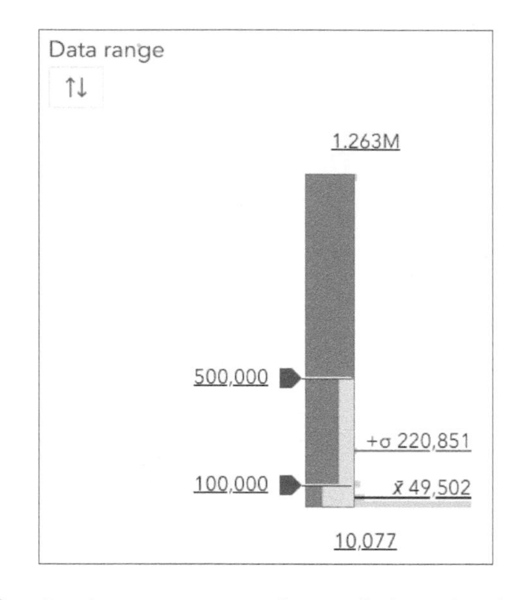

5. On the map, review the symbology for the cities.

   The symbols for small cities are difficult to see and Dublin is too large, so you'll adjust the size range of the symbols.

6. In the Styles pane, under Size range, change the lower size range number to **8** and the upper size range number to **22**.

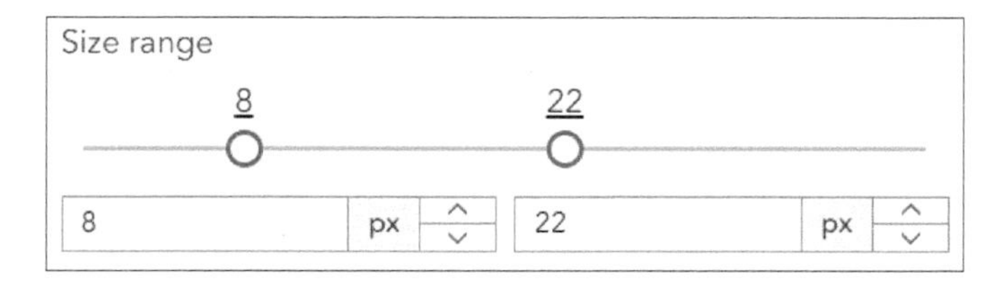

7. In the Style options pane, click Done.

Duplicate the Cities layer, apply a filter, and set visible ranges

Now you'll design functionality that's specific to interactive web maps. You'll duplicate the Cities layer and filter the duplicate layer so that only the major cities are shown. Then you'll set the visible ranges for the two layers. The layer with the major cities will appear when users zoom out on the map. When they zoom in, the layer with the major cities will turn off, and the layer with all the cities will appear. This effect makes more cities visible when they're needed.

1. In the Layers pane, on the Cities layer, click Options and click Duplicate.
2. Rename the new layer **Major Cities.** Click OK.
3. Turn off visibility for the original Cities layer.

   You'll filter the Major Cities layer so that only the top three most-populous cities are shown.

4. With the Major Cities layer selected, on the Settings toolbar, click Filter.
5. In the Filter pane, click Add expression.

   An expression will limit the visible features. Expressions are faster and simpler than editing and deleting features that you don't want to display.

6. Under Expression, create the following expression: **Population is greater than 100000.**

   Expression

   Population

   is greater than

   100000

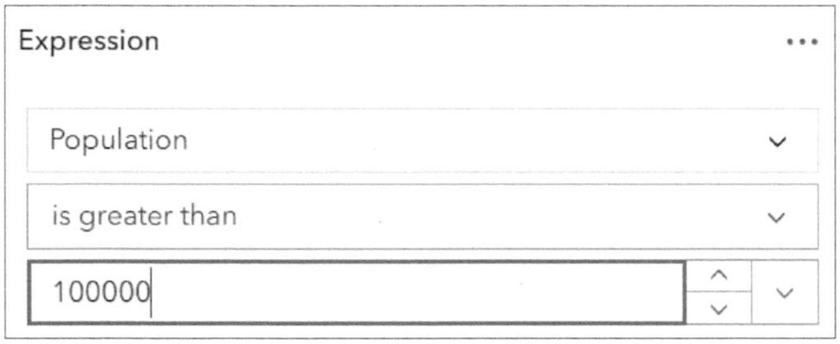

   On the map, all the cities with a population of fewer than 100,000 are uneditable (gray). You can see the filter change the map on-the-fly. When you click Save, the uneditable cities are filtered completely.

7. Click Save.

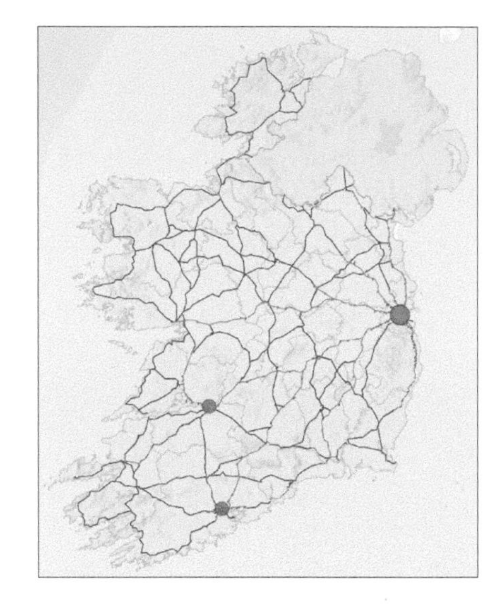

Now you'll set the visibility range for each layer.

8. On the Settings toolbar, click Properties.

9. In the Properties pane, under Visibility, click Room. From the list, click State/Province 1:3,000,000.

The adjusted visible range is shown.

> *The black triangle indicates the current scale of your map, which may vary from what's shown in this image.*

Visibility

Visible range

World State

You have set the visibility range so that the Major Cities layer displays when users zoom out on the map. The layer continues to display until users zoom in closer than the State scale. At that point, the layer turns off.

10. On the map, zoom in and out to see the major cities appear and disappear at different scales. When you've finished, zoom to the first scale at which the major cities disappear.

> *The major cities shouldn't be visible on the map.*

11. Select the Cities layer and turn on its visibility.

12. In the Properties pane, under Visibility, click World. From the list, click State/Province 1:3,000,000.

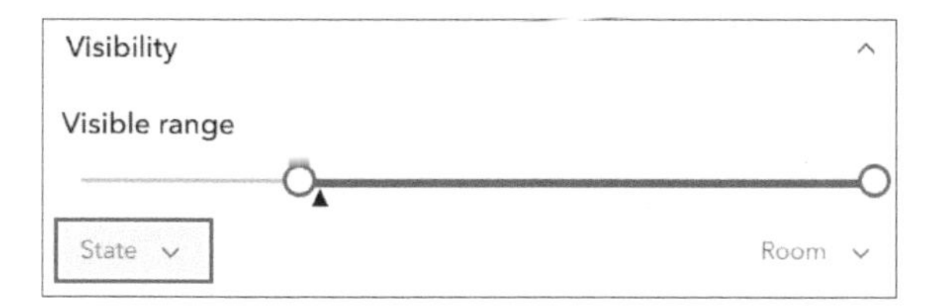

13. Zoom in and out to see the visibility for the two layers toggle on and off. Watch the Layers pane and notice that each layer's name is uneditable (gray) when it's no longer visible because of the visible range adjustments you've made.

On your own

Set the visible range of the Cities layer to turn off when users zoom in beyond the Cities scale. To the right of the range, set the level to Cities 1:160,000. A city is a polygon (it has no exact point), so seeing a point on the map when you're zoomed to a particular city is confusing. This range setting turns off the City point once you've zoomed in far enough.

When users zoom out, the Major Cities, Counties, and Roads layers become unnecessary and confusing. Choose a visible range at which to turn them off.

Add labels

Now that you have the map's features configured logically, it's time to add labels.

1. Zoom in until the city points in the Cities layer are visible.

2. With the Cities layer selected, on the Settings toolbar, click Labels.

3. In the Label features pane, click Add label class.

   By default, the City field is being used for the labels. Its visible range is the same as the Cities layer. Next, you'll choose a label style to help viewers associate the labels with the cities. You'll also apply a halo around each label that's the same color as the Ireland layer. A halo helps label text to stand out on a map.

4. In the Label features pane, click Edit label style.

5. In the Label style pane, under Font, click the down arrow, and choose Josefin Slab Bold.

6. Under Size, increase the size to **15**.

7. Under Color, click the default and enter the hexadecimal value **776491**. Click Done.

8. Under Halo, click the default color and enter the hexadecimal value **e2eed4**. Click Done.

9. Under Size, increase the size to **3**.

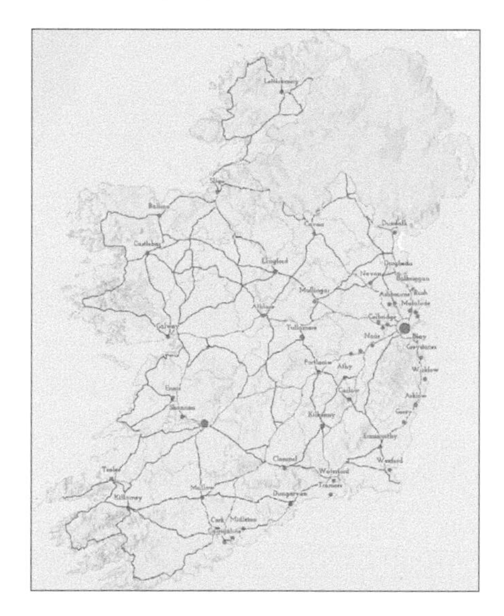

On your own

Apply the same label settings to the Major Cities layer. When you've finished, close the Label style pane.

Add an effect

Effects can highlight specific features or an entire layer. Here, you'll apply an effect to the Ireland layer to enhance its appearance.

1. In the Layers pane, select Ireland.
2. On the Settings toolbar, click Effects.
3. In the Effects pane, turn on Drop Shadow.
4. In the Drop Shadow pane, set the Width to **2**.
5. In the Offset area, set the X-offset to **3** and the Y-offset to **4**.
6. Set the Opacity to **70**.
7. Close the Drop shadow pane.

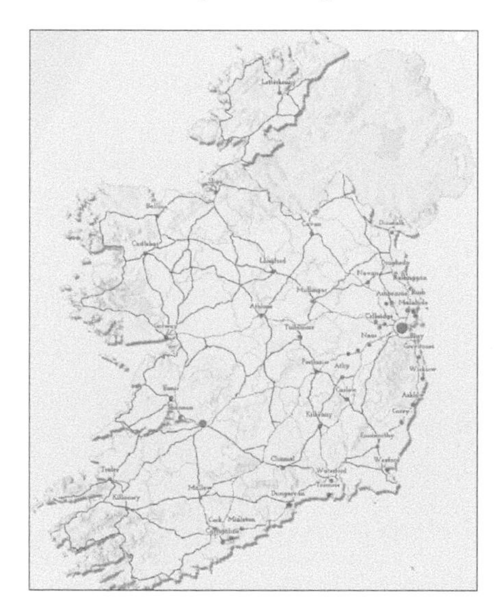

Finalize the legend

To help users understand your map, you'll choose which layers appear in the legend and make final adjustments.

1. On the Contents toolbar, click Legend.

   The legend shows the layers that will be listed. Northern Ireland is included on the map for context, so listing it isn't necessary. You'll remove it from the legend.

2. On the Settings toolbar, click Properties.
3. At the top, change the heading selection to the Northern Ireland layer.
4. In the Properties pane, under Symbology, turn off the toggle next to Show in map legend.

Save and share the map

1. On the Contents toolbar, click Save and open > Save as.
2. Save your map with the following settings:

   - Title: **Ireland <your initials>**
   - Folder: Top 20 Tutorial Content
   - Tags: **Top 20**, **Cartography**, **Ireland**
   - Summary: **A map of Ireland.**

   > *If you need to create the folder, click the down arrow and choose Create new folder. Type* **Top 20 Tutorial Content**.

3. Click Save.
4. On the Contents toolbar, click Share map.
5. In the Share pane, click Everyone (public) and click Save.

Take the next step

Add a layer of urban areas to the map. Find the layer by searching **owner:Top20EssentialSkillsForArcGISOnline Urban**. Order the layer in the list as you think appropriate. Filter it to display large urban areas using the Shape_Area field. Choose an area size that you think works best. Set the visibility scale, and style the layer with an appropriate color. Configure the pop-ups using skills you learned earlier in the book.

Summary

In this chapter, you created an interactive map with layers symbolized logically, showcasing the core functionality of a web map.

Workflow

1. Add all necessary layers.
2. Rename layers and order them in a logical hierarchy.
3. Apply styles, starting from the bottom:
    - Blend
    - Filter
    - Group
4. Set visibility ranges.
5. Add labels and effects.

CHAPTER 6

Creating a scene and exploring 3D visualization

Objectives

- Explore line of sight.
- Create a dimensions layer.
- Add a spatial filter.
- Calculate an elevation profile.
- Add effects.
- Create a a local scene.
- Add 3D layers.
- Configure content and style layers.
- Capture slides.

Introduction

In this chapter, you'll create a scene and take your mapping to the third dimension. You'll learn how 3D adds greater context by providing realistic visualization and advanced analytics.

Scenes (3D maps) can be categorized into two general types. First is a reality-based view in which you create an immersive style that mimics the real world. This type of scene contains many elements (see figure 6.1), and a variety of analyses can be performed on these real-world objects.

The second approach is to focus on the 3D visualization of attributes. This approach doesn't represent the real-world look and feel of a cityscape, but it's an effective way to visualize data (see figure 6.2).

Figure 6.1. A scene of Montreal, featuring numerous elements, including clouds, sky, hills, skyscrapers, and other built structures.

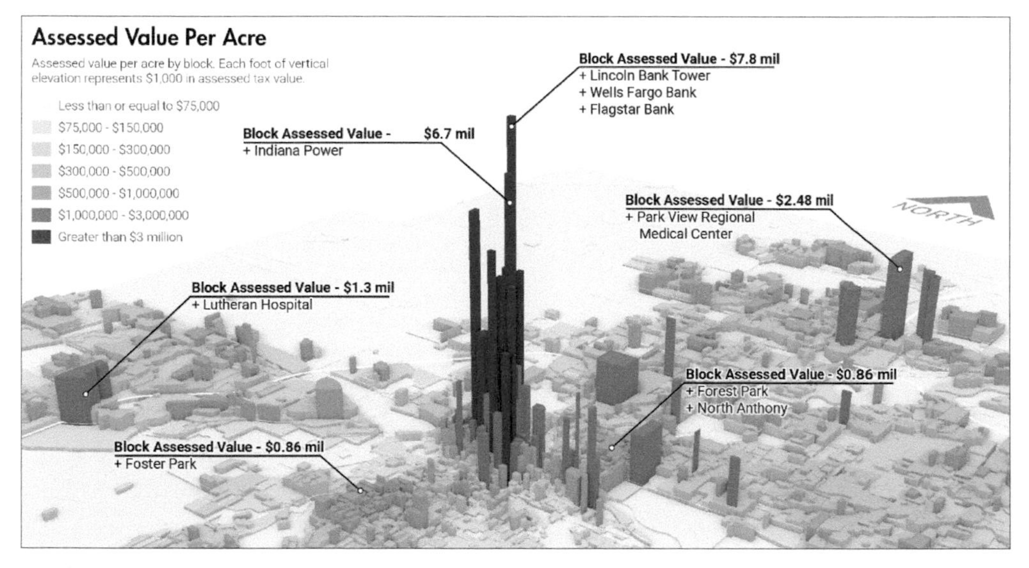

Figure 6.2. Tax values are shown based on acreage by block in Fort Wayne, Indiana. Extrusion is used to generate 3D representations of dollar amounts–the higher the block, the higher the assessed value of that plot.

Depending on what you're trying to accomplish, you can decide which approach is best for your needs; there are no rules, so you may even find creative ways to mix and match these approaches!

Tutorial 6-1: Tour a scene and make it your own

In this tutorial, you'll open an existing scene and explore line of sight, add a layer to show the real-world dimensions of objects, remove features, calculate an elevation, add weather effects, and save a slide.

Sign in and open a scene

You'll start by exploring an existing scene.

1. Sign in to ArcGIS Online.
2. On the top navigation bar, click the magnifying glass to search for items in ArcGIS Online.
3. Type **owner:Top20EssentialSkillsForArcGISOnline Montreal** and press Enter.
4. On the left, under Filters, turn off the filter for your organization and expand your search to all of ArcGIS Online.
5. On the Montreal 3D item, click Open in Scene Viewer.

   A 3D scene layer of Montreal, Canada, is overlayed on the Imagery basemap. This overlay provides a detailed 3D representation of the city.

6. Zoom and pan to explore the city.

View a slide and explore line of sight

1. Point to the center bottom of the screen to see the Line of Sight slide. Click Line of Sight.

Each slide takes you to a viewpoint, similar to a bookmark in a 2D map. You can use the Line of Sight tool to visualize what you would be able to see from a particular vantage point. Once you choose an observation point and a target, green lines are added to the scene, showing unobstructed lines of sight. Red lines indicate a visual obstruction between observation points and targets.

Certain windows in the taller building to the left probably have their views blocked by the smaller building to the right. You'll calculate line of sight to see how views from the larger building are affected by the smaller building.

2. In the lower left, click the arrows to expand the Designer toolbar.

3. On the Designer toolbar, click Add layers > Line of sight.

   The Line of sight pane opens.

4. In the scene, click one of the windows about halfway up the right side of the taller building to add an observation point.
5. Extend the line of sight to the green, open area and click to add a point.

The first stretch of the line will be green but will become red, showing where the clear line of sight ends.

6. Click to add a few more lines of sight that are unobstructed, and add a few to show where the building begins to obstruct the view.
7. Double-click when you have finished adding lines of sight.

8. Drag the orange observer point to different window locations on the taller building. Move it to a window with a clear line of sight for all the lines you've added.
9. In the Line of sight pane, click Done.

Create a dimensions layer

You can add a dimensions layer to measure the dimensions of specific objects in a scene. Here you'll create a simple view that can show city planners the height of the obstructing building.

1. In the bottom center of the screen, click the Dimensions slide.

   To help viewers focus on the dimensions of the building, the Dimensions slide was saved with the Streets (night) basemap, which provides a more focused view of the buildings.

2. In the Layer Manager pane, for the Montreal, Canada Buildings layer, click the options button and click Layer style.
3. Under Choose a drawing style, click Options.

4. Next to Texture, click the symbol and click Color only.

   There are a number of options here to adjust the style, but the default provides a simplified schematic view.

5. On the Designer toolbar, click Add layers > Dimensions.

6. Click the top corner of the building facing you and click the corner of the building directly below it on the ground.

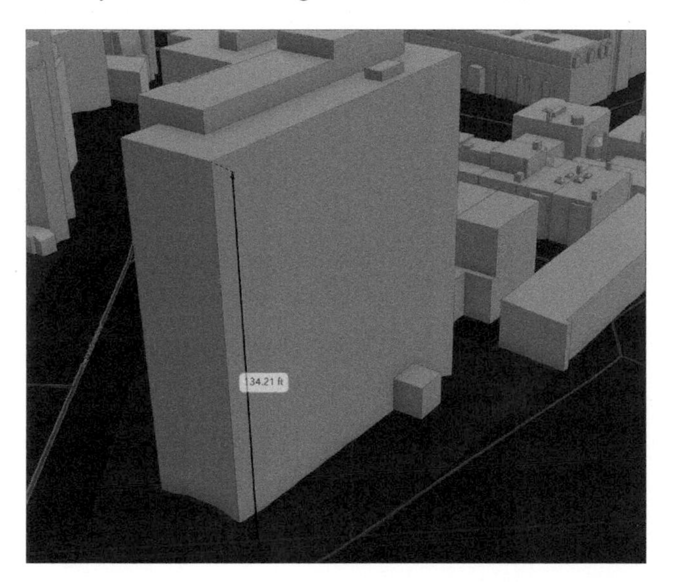

7. In the Dimensions pane, click the color picker for Line, change the color to red, and click Done.

8. Click the color to Label, choose red, and click Done.

9. For Size, click the down arrow, and choose large.

10. On the Dimensions pane, click Done.

    Now your scene includes a layer showing the dimensions of the building. This makes it easy to communicate important details to others.

On your own

Reopen the Montreal, Canada Buildings Layer style pane and change the drawing style back to show the original texture instead of the solid color. Click Done twice.

Add a spatial filter

Next, you'll add a spatial filter to remove a feature. In this case, you'll remove the obstructing building.

1. On the bottom center of the screen, click the Spatial Filtering slide.
2. In the Layer Manager pane, for the Montreal, Canada Buildings layer, click the options button, and click Layer properties.
3. In the pane, under Spatial filters, click Configure spatial filters.

   You can draw an area to create a filter, or click features interactively, which is the method you'll use next.

4. Under Filter by selection, click Select features to exclude.
5. Click to highlight the obstructing building to the left of the large building.

6. Click Done and click Done.

   The building is removed from the scene.

Calculate elevation profile

Numerous tools are available for 3D analysis. You'll analyze the elevation profile of a road.

1. On the bottom center of the screen, click the Elevation slide.
2. In the upper right of the scene, hover over each tool to see its name. Click Scene tools.
3. Click the Elevation Profile tool.

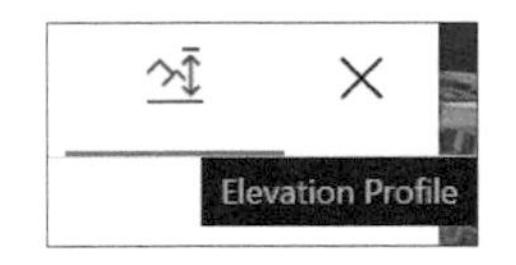

4. Orient the scene so that you're looking straight down, and the parking lot and the tower at the top of the hill are visible.
5. Click inside the parking lot and trace the road to the tower. Double-click to finish.

The Elevation Profile tool calculates key information about the route.

6. In the Elevation Profile tool pane, for Ground, expand the arrow to see the route details.

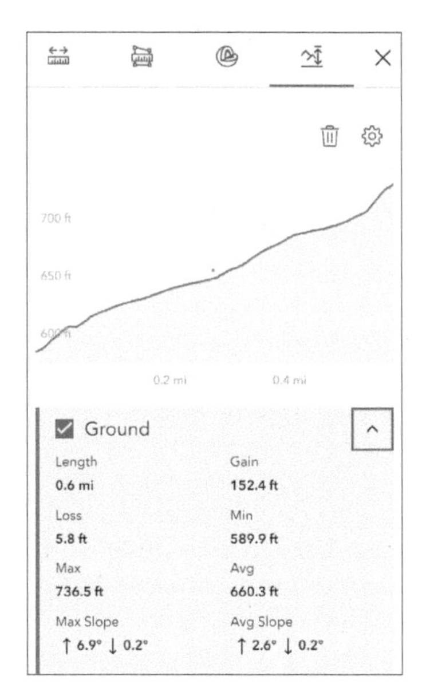

7. Close the Elevation Profile tool.

Add daylight and weather effects

Now you get to have some fun adding visual effects.

1. On the bottom center of the screen, click the Effects slide.
2. In the upper right of the scene, find and open the Daylight/Weather tool.
3. For Date, choose November 15.
4. Adjust the slider to 3:30 pm.
5. Confirm that the Shadows check box is checked.
6. At the top, click the Weather tab.
7. Click the button for Rainy.
8. Adjust the Cloud cover and Precipitation sliders to your liking.

Capture a slide

Next, you'll choose a viewpoint and capture a slide that you can share with others.

1. Pan and zoom to a location and viewpoint that you find interesting.
2. On the Designer toolbar, click Slide Manager.
3. Click Capture slide.
4. Give your slide a clear, descriptive name.

> **Hint:** If you want to screenshare with others and show the scene without the toolbars and menus, you can click Present on the navigation bar at the upper right of the window. To return to the original view, you'd click Exit presentation.

Save and share your scene

1. On the Designer toolbar, click Save.
2. Add the following details:
    - Title: **Montreal in 3D**
    - Summary: **A scene showing Montreal, Canada.**
    - Tags: **Top 20, Montreal.**
3. Click Save.

4. In the upper right of the scene, click the share button, which is second from the bottom.
5. Click Change Share Settings.
6. On the Item page, under Share, click Edit and click Everyone (public).
7. Click Save.

Tutorial 6-2: Build and style a scene

As you've learned, data is often provided in 2D, but sometimes it's easier to understand and visually more interesting to show it in 3D. This tutorial gives you the basics for how to build your own scene.

Open a new scene

You'll start with a new scene and learn how to choose between a global or a local scene.

1. On the top navigation bar, click Scene to open Scene Viewer.
2. In the gallery, click New scene.

   You can choose between a global scene or a local scene. A global scene shows the whole world as a globe. This is a good choice if you're working with global data, but because it calculates for the curvature of the earth, it can be unnecessarily complex if you're only working in a relatively small area. A local scene uses a planar surface, so it draws much quicker and is typically suitable for a city or regional extent. (A planar coordinate system is a 2D measurement system that locates features on a flat plane.)

3. On the right of the top navigation bar, click New Scene.
4. Choose New Local Scene.

   The scene changes from a globe to a planar surface.

5. In the upper right, search for **Kew Gardens, England** and press Enter.
6. On the left of the Designer toolbar, click Add layers and click Browse layers.

> **Hint:** If you want to see the whole Designer toolbar, click the Expand button in the lower left.

7. Click My content and change the selection to ArcGIS Online.

   You can now search for any public layer in ArcGIS Online.

8. In the search box, type **owner:Top20EssentialSkillsForArcGISOnline trees**.
9. On the Kew Trees layer, click Add.

   This hosted feature layer is a 2D layer that you have moved into a 3D environment. Points are placed on the scene's 3D surface.

10. In the Browse layers pane, click Done.
11. Zoom and pan around the scene so that the points fill the view.

    These points are from a fictional tree survey with attributes that include tree species and height, which will be used to create 3D trees near Kew Gardens. A rotation attribute is also included, though in this case it's less important and will be used for visual purposes only.

Choose a basemap

Before you continue, you'll choose a basemap that will help provide a sense of the area.

1. In the upper right of the scene, find and click the Basemap tool.
2. Click the down arrow, and from the list, click the Imagery basemap.

   Now you can see the existing trees and how the point symbols of the fictional trees will fit into the landscape.

3. Switch to the Streets 3D basemap and close the Basemap selection pane.

   The Streets basemap provides context for the trees you'll add.

Symbolize points in 3D

The layer is symbolized with points, but it's not clear that they represent trees. So you'll turn the points into 3D tree symbols.

1. In the Layer Manager pane, click the Kew Trees layer.
2. Under Choose the main attribute to visualize, click the down arrow and choose Species.

   Default colors are applied to the symbols, differentiating the species.

3. Under 3D Types, click Select.

The simple 2D points are changed into default 3D symbols. You'll choose more meaningful 3D symbols that will reflect each tree species.

Set the 3D symbol based on the species

To assign more meaningful symbols, you'll use the layer's attributes to drive the 3D symbology. Under Attribute Values, four species of trees are listed, each with a default color and symbol.

1. Under 3D Object, click Options.
2. Confirm that the symbol for Alder Buckthorn is selected.

3. Below the list of symbols, click Marker.

4. In the window, click the down arrow. At the bottom of the list, click Vegetation. (Do not click Schematic Vegetation.)

5. Click the first symbol, Alder Buckthorn.

6. Click Done.

On your own

Repeat this process for the remaining symbols, pointing to the symbols to find the correct species.

Set the symbol size based on the height attribute

The layer has an attribute for height, so you'll use that to set the height of each 3D symbol.

1. For Size, click the down arrow and choose Height.

Each symbol's height has been adjusted.

2. Zoom and pan around the scene to review the changes you've made.

On your own

Each tree symbol has the same orientation, making each one look a bit too similar. Try using the Rotation field to add variation.

3. When you've finished adjusting the layer settings, click Done and click Done.

Clip to extent, save, and share

In a local scene, you can clip the basemap and layers to the current extent of your view. Clipping is useful for improving performance. Clipping also focuses the scene so that people don't inadvertently navigate away from it.

1. Orient the scene so that you're looking straight down, and zoom out to show some of the larger context.

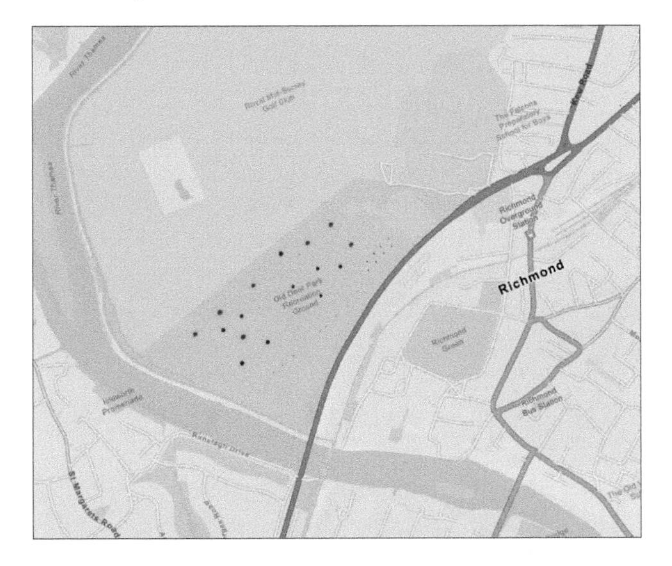

2. On the Designer toolbar, click Properties.
3. In the Properties pane, click Clip to Extent.
4. Zoom out to see the newly clipped boundary.

On your own

On the Properties pane, change the Background color to something you think looks good. Use the Daylight/Weather settings and add shadows. Find an interesting viewpoint and capture a slide. If you need, review the steps that described how to do this in the first part of this chapter.

5. On the Designer toolbar, click Save and add the following settings:
   - Title: **Kew Gardens scene**
   - Summary: **3D tree design**
   - Tags: **Top 20, Kew**
6. Click Save.
7. In the upper right of the scene, click the button that is second from the bottom (the share button), as you did earlier.
8. Click Change Share Settings.
9. On the Item details page, under Share, click Edit, and share the scene with Everyone (public).

Take the next step

You can turn your scene into a 360-degree virtual reality (VR) experience in just a few clicks.

1. Open one of the scenes you created in Scene Viewer.
2. On the Designer toolbar, click Create app > 360 VR Experience.

   A new tab opens in your browser.
3. Confirm the name and click Publish. Wait for the publishing process to complete, which may take several minutes.
4. Click Open in 360 VR.
5. Explore the viewpoints.

Summary

In this chapter, you explored line of sight, created a dimensions layer, removed a feature from the scene, calculated an elevation profile, added effects, captured a slide, and applied 3D symbology based on attributes from a layer.

Workflow

1. Choose to create a global or a local scene.
2. Build on a basemap that provides context for the content.
3. Add layers.
4. Configure content and style your layers to tell your story.
5. Capture slides that show interesting viewpoints.
6. Save and share.

CHAPTER 7

Using ArcGIS Living Atlas of the World

Objectives

In this chapter, you'll learn about ArcGIS Living Atlas of the World from two user experiences: the ArcGIS Living Atlas website and Map Viewer.

- Learn about ArcGIS Living Atlas and its role in ArcGIS.
- Discover the different types of content that ArcGIS Living Atlas offers.
- Learn about the ArcGIS Living Atlas website and resources.
- Search for different kinds of content in ArcGIS Living Atlas.
- Make maps using the ArcGIS Living Atlas website.
- Make maps using ArcGIS Living Atlas in Map Viewer.
- Style and filter ArcGIS Living Atlas layers in a web map.
- Save layers for future use.

Introduction

Content is the foundation of our work. As GIS users, we all need access to useful and reliable data. ArcGIS Living Atlas is the foremost collection of worldwide geographic information, containing basemaps, layers, maps, apps, and tools from the global GIS community and Esri®.

The content in ArcGIS Living Atlas is authoritative, curated, and continually updated, forming a vast collection of essential maps, layers, and more. When we think about what's in ArcGIS Living Atlas, the word content is typically used. It's a distinction made because it's not just raw data that needs to be manipulated or refined—it's ready-to-use content that's been crafted for the best experience across a variety of workflows.

ArcGIS Living Atlas has a website (livingatlas.arcgis.com), a destination where anyone can explore its content. ArcGIS Living Atlas is also deeply embedded in the ArcGIS ecosystem, an essential part of how we use and make maps and apps.

In this chapter, you'll learn how to use the ArcGIS Living Atlas website and discover content. You'll use Map Viewer to make maps and learn how to filter, style, and save what you use for other maps.

Tutorial 7-1: Explore the ArcGIS Living Atlas website and make a map

The ArcGIS Living Atlas website is a destination for news, learning, discovering content, and community participation. The site is arranged into tabs found at the top of the page. In this section, you'll explore several of the tabs.

Explore the Home tab

1. In a browser, go to the ArcGIS Living Atlas home page at **livingatlas.arcgis.com**.
2. At the top of the page, click Sign In and sign in to your ArcGIS Online account.

   Anyone can visit the site and begin exploring, but when you sign in to your ArcGIS Online account, you can use the site to support your workflows. You can use the search and filter capabilities to find layers to start your map, and you can designate items as favorites to make them easy to find later. Some ArcGIS Living Atlas content is available only to subscribers with an ArcGIS Online account, so you won't be able to view this content without signing in first.

   The Home tab includes search functionality and information about what's new in ArcGIS Living Atlas.

3. Click the Home tab, find a topic of interest, and view more details.
4. When you've finished exploring, go back to the ArcGIS Living Atlas home tab.

Explore the Apps tab

Content from ArcGIS Living Atlas is used to make a variety of ready-to-use apps for visualization and analysis.

1. On the navigation bar, click the Apps tab.
2. Scroll down to find Air Quality Aware and click the thumbnail or title.
3. After reviewing the app summary, click View Application.
4. Explore the map. Click different locations to learn more about current air quality conditions.

   The side panel displays current and forecast conditions, along with wind speed and population demographics. If you click a monitoring station (the circle symbols on the map), you'll see additional information from the station in the pop-up.

   > **Hint:** Click the question mark in the upper right to learn more about the app and the ArcGIS Living Atlas content it uses.

On your own

Explore other apps that you find of interest. Note that one of the apps is the Live Feeds Status app, a dashboard listing live feeds and their usage trends, updates, and more information. You'll use live feeds later when making a map using ArcGIS Living Atlas content in Map Viewer.

5. When you've finished exploring, return to the home tab.

Explore the Browse tab and review a layer

The Browse tab allows you to discover content using keywords and to filter by type, category, region, and even date (past month, year, or custom date).

> **Hint:** You can also browse ArcGIS Living Atlas content by using the search box at the top of the home page.

You'll create a map showing median household income and unemployment statistics for counties in Texas using the latest demographic layers from American Community Survey (ACS) and Bureau of Labor Statistics (BLS).

1. Click the Browse tab.

   You'll see a search bar and other options for filtering content.

2. In the search box, type **household income** and press Enter.

   > **Hint:** Under the search box, click Search Examples to see other types of searches you can perform.

   Your search returns many results, including content from other countries, different vintages, and different types of items: layers, dashboards, stories, maps, scenes, and tools. You'll refine the search to return only recent, authoritative layers for the United States.

   Directly below the search box, you'll see a number of categories with distinct icons. These categories allow you to tailor your search.

3. Click the People category down arrow and check the Income check box.

   People
   - [ ] Population
   - [ ] Housing
   - [ ] Neighborhoods
   - [ ] Jobs
   - [x] Income
   - [ ] Spending
   - [ ] Health
   - [ ] Education
   - [ ] At Risk
   - [ ] Public Safety

4. Under the category choices, click All content types and choose Layers.

All content types
- All content types

Maps (30)

Layers (24)

> *As new layers are added to ArcGIS Living Atlas, the number of layers available may vary from what's shown in the preceding image.*

5. Click All time and choose Past Year.
6. Click All regions and choose United States.
7. Check the Authoritative-only content check box.

   The applied filters are shown.

Hint: If you want to clear a filter, click the X and uncheck the Authoritative-only content check box.

8. In the results, find ACS Median Household Income Variables – Boundaries and click the thumbnail or title to view the item details, which open in a new browser tab.

   The description for each ArcGIS Living Atlas item provides information about the source, the date of the most recent update, and so on.

   You can use Add to Favorites on the item details page to make content easier to find later. You can also set the item as a favorite directly from the item card in your search results without having to open the item details.

9. Click Add to Favorites beneath the thumbnail to add this item to your favorites.
10. Click the thumbnail to open it in Map Viewer.

Hint: You can open the layer in Map Viewer from the options menu (ellipses) on the item card in your search results without viewing the item details.

Choose the basemap and layers

Like with any other layer, you can configure ArcGIS Living Atlas layers and apply filters and effects.

If your default basemap is Topographic, you'll notice that the basemap labels are hidden by the household income layer you just added. You'll choose a different basemap with a reference layer that will place the labels on top of the layer.

> *Review the basics for navigating the user interface in Map Viewer. The Contents toolbar is the dark vertical toolbar on the left, and the Settings toolbar is the light vertical toolbar on the right. When you click or select something on a toolbar, a pane usually opens, allowing you to adjust a related setting. Sometimes, clicking something in a pane opens an additional pane that shows more settings.*

1. On the Contents toolbar, click Basemap and try a few basemaps. Choose one that doesn't compete with the layer and includes a reference layer that places the labels on top. If it's available, Human Geography Map or Light Gray Canvas is a good option.

   > *The images in this tutorial use Human Geography Map as the basemap.*

2. On the Contents toolbar, click Layers.
3. In the Layers pane, expand the ACS Median Household Income Variables – Boundaries group layer by clicking the Expand button in front of the name.

The group layer contains three layers—State, County, and Tract. Each layer has a visible range applied. This setting turns visibility on or off, depending on how far the map is zoomed in.

4. Zoom in and explore the map, view the different layers, and click features to open the pop-ups displaying median household income data.

   To focus on only the counties, you'll turn off layer visibility for the State and Tract layers in the group layer.

   There are two similar-sounding settings related to seeing a layer on the map: visible range and visibility. Visible range determines whether a layer is visible at a certain scale (based on how far you are zoomed in). Check the Layers pane to see whether a layer is out of range. If it's out of range, its name appears in light gray and the layer won't be visible on the map because it's been set to be invisible at the current scale.

   The other setting, layer visibility, determines whether a layer can be seen at any scale or range. Turning off layer visibility prevents it from being seen at any scale. Layer visibility is controlled by the Visibility button (eye icon) in the Layers pane to the right of each layer name (near the options button).

5. In the Layers pane, point to the State layer and click the Visibility button (eye icon) to turn off layer visibility.

6. Turn off layer visibility for the Tract layer.

7. Zoom in and out.

   The State and Tract layers no longer display.

8. In the Layers pane, click the County layer to select it.

9. In the Properties pane, under Visibility, drag the Visible range circles to either side, making the layer visible at all ranges.

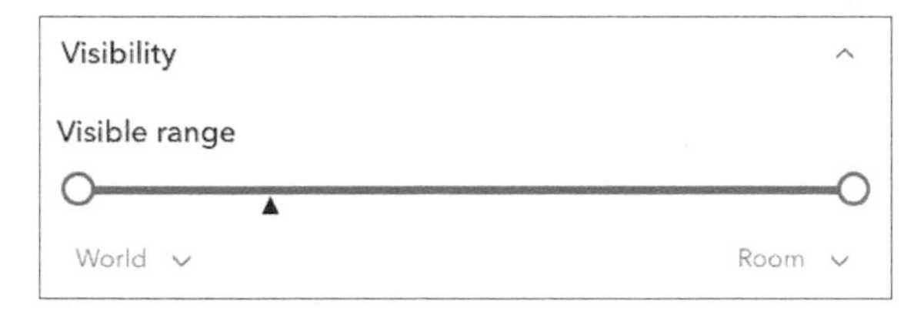

10. Zoom in and out on the map.

    The County layer is now visible at all scales.

Add a filter

Many layers in ArcGIS Living Atlas are nationwide layers, including the ACS Median Household Income Variables – Boundaries layer. If you want to limit the layer to display a certain area, you can use attribute filters.

1. On the Settings toolbar, click Filter.
2. In the Filter pane, click Add expression.
3. Use the down arrows and choices to create the following expression: **State is Texas.**

Expression

State

is

Texas

4. Click Save, and pan to Texas.

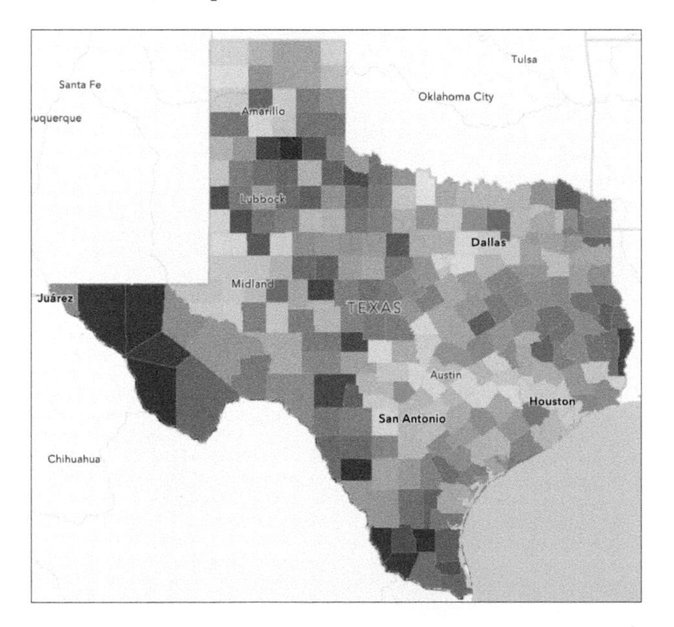

Hint: If you need to remove a filter, click Remove filter in the Filter pane.

Save and share your map

Now that you've finished making changes to the ArcGIS Living Atlas layer settings, you can save the map. Saving keeps the changes you've made but still references the original ArcGIS Living Atlas source so that you're always up to date.

1. On the Contents toolbar, click Save and open > Save as.
2. In the Save map window, add the following settings:
    - Title: **Texas Income <your initials>**
    - Folder: Top 20 Tutorial Content
    - Tags: **Top 20, ArcGIS Online, Texas**
    - Summary: **Texas income from ArcGIS Living Atlas.**

    > *If you need to create the folder, click the down arrow and choose Create new folder. Type* **Top 20 Tutorial Content**.

3. Click Save.
4. On the Contents toolbar, click Share map.
5. In the Share pane, click Everyone (public) and click Save.

Tutorial 7-2: Create a map using Map Viewer and ArcGIS Living Atlas

In this section, you'll create a map using ArcGIS Living Atlas content. You'll start from Map Viewer with a new empty map and search for ArcGIS Living Atlas layers using Map Viewer.

Your map will show current drought conditions with live feeds of current wildfires and air quality.

Browse ArcGIS Living Atlas from Map Viewer and add a layer

1. If you signed out of ArcGIS Online, sign in again.
2. On the Contents toolbar, click Save and open > New map.

    A new map appears. The Layers pane is open by default.

3. In the Layers pane, click Add.

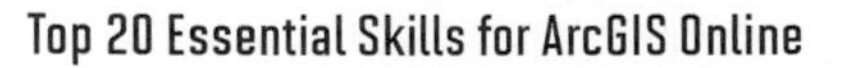

4. In the Add layer pane, click My content, and click Living Atlas.

   The collection of layers appears.

   > **Hint:** Click the filter button next to the search box to view filtering options. Many of these options are similar to the filters you used when browsing for content on the ArcGIS Living Atlas website. These filters are useful but not necessary to complete the map. Close the filter.

5. In the search box, type **drought**.

   The list of layers updates to show relevant results.

6. Find USA Drought Intensity – Current Conditions. If you have trouble finding it, enter the complete layer name to narrow the search.

7. Click the item card to view a summary of the layer information, and read the summary.

8. Close the pane to return to the search results.

9. Click Add to add the layer to your map.

   > **Hint:** Once the layer is added to your map, the button changes to Remove for easy removal.

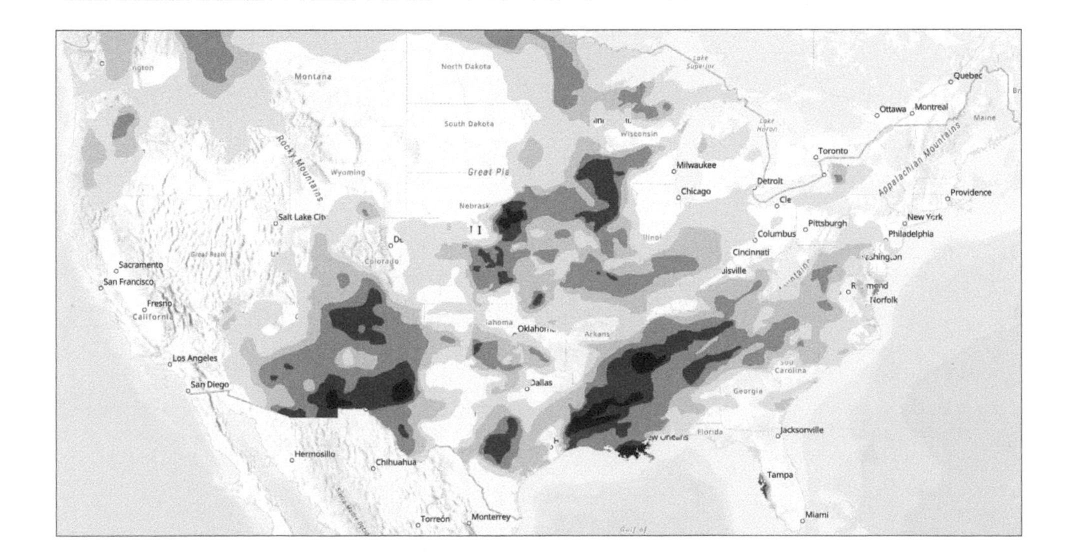

> *As drought conditions change over time, the layer you're using will be updated and may appear different from this image.*

If your default basemap is Topographic, the basemap labels will be hidden by the drought layer. You'll choose a different basemap with a reference layer that will place the labels on top of the layer.

Blend the drought layer with the basemap

1. On the Contents toolbar, click Basemap and choose Terrain with Labels.

   Terrain with Labels is a good choice because it doesn't compete cartographically with the layer and includes a reference layer that places the labels on top.

2. Zoom to an area of drought and examine the map.

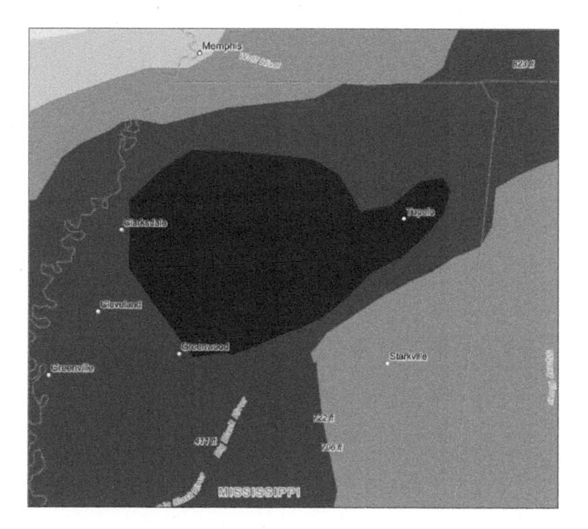

   Even though the labels are visible, the terrain below the drought layer is not.

3. On the Contents toolbar, click Layers. Ensure that the USA Drought Intensity – Current Conditions layer is selected (indicated by a vertical blue bar).
4. On the Settings toolbar, click Properties.
5. In the Properties pane, expand Appearance.
6. Under Blending, click the arrow next to Normal and click Multiply. Close the pane.

   The basemap terrain is now visible.

Multiply emphasizes the darkest parts of overlapping layers by multiplying colors of the top layer and the background layer. You can also adjust the layer transparency to achieve the desired visual result.

7. Zoom out to see the entire United States.
8. In the Layers pane, click Add.
9. From the Add menu, choose Living Atlas.
10. In the search box, type **live feeds** and press Enter.

    The layer list shows the live feed collection. ArcGIS Living Atlas includes a valuable collection of live feeds that show current conditions. These feeds are from organizations such as the United States Geological Survey (USGS), the National Oceanic and Atmospheric Administration (NOAA), the National Weather Service (NWS), and others, and describe conditions such as recent earthquakes, storms, wildfires, flooding, stream gauge levels, and more. Live feeds are updated regularly, some as often as every few minutes.

    > **Hint:** To view a list of live feeds with current trends, usage, and updated information, go to links.esri.com/LiveFeeds.

11. Find live layers that you consider interesting. Add several live layers to see what they look like.
12. Scroll down in the search results to locate USA Current Wildfires.
13. Add the layer to your map so that drought and fire conditions are shown.

Save and share your map

1. You'll save your map showing live drought and fire conditions.
2. On the Contents toolbar, click Save and open > Save as.
3. In the Save map window, add the following settings:
    - Title: **Drought and fire conditions <your initials>**
    - Folder: **Top 20 Tutorial Content**
    - Tags: **Top 20, Drought, Fire**
    - Summary: **Live drought and fire conditions from ArcGIS Living Atlas.**

4. Click Save.

5. On the Contents toolbar, click Share map.

6. In the Share pane, click Everyone (public) and click Save.

Take the next step

Apply a filter to a layer to filter certain attributes. Expand the USA Current Wildfires group layer and select the Current Incidents layer. Add a filter that omits prescribed burns. Use effects to create a drop shadow to highlight the fires.

Summary

In this chapter, you learned how to search ArcGIS Living Atlas and use its content in different workflows. You now know how to:

- See what's new at the ArcGIS Living Atlas website.
- Find and use ArcGIS Living Atlas apps.
- Use the website browse capabilities to filter and find content you want.
- Use ArcGIS Living Atlas search capabilities in Map Viewer to find and add layers to your map.

CHAPTER 8

Locating things and places

Objectives

- Learn how to find features by attribute and location.
- Build a compound query.
- Generate travel-area polygons based on travel time.
- Summarize statistics within a polygon.

Introduction

In this chapter, you'll use tools to accomplish common tasks that people ask of maps. The first tool you'll try, Find by Attributes and Location, can find things based on what and where they are. The other tool, Generate Travel Areas, lets you identify the areas you can travel to in a given amount of time.

Tutorial 8-1: Find by attribute and location

You've probably used a mapping app to perform various map-related functions, such as finding nearby fuel stations or restaurants. This calculation may seem simple, but it's really asking for two calculations to be made: an attribute and a location. The attribute calculation asks for points that have been labeled as gas stations. The location calculation asks to show points in a relevant area.

In this tutorial, you'll map fuel stations that are located within 1,000 meters of toll plazas. To do this, you'll run the Find by Attributes and Location tool, which allows you to combine your request for attributes and location in one search.

Sign in and open a map

First, you'll set up your workspace.

1. Sign in to ArcGIS Online.
2. On the top navigation bar, click the magnifying glass to search for items in ArcGIS Online.
3. Type **owner:Top20EssentialSkillsForArcGISOnline fuel stations** and press Enter.
4. Under Filters, turn off the filter for your organization and expand your search to all of ArcGIS Online.
5. On the Find Fuel Stations item in the search results, click Open in Map Viewer.

   The Find Fuel Stations map opens with two layers—Rural Facilities and West Bengal Toll Plazas.

Explore the map

Next, you'll get a better sense of the layers you're working with.

1. Click some points from each layer and read the pop-ups.
2. Click a point in the Rural Facilities layer to open a pop-up. Find the Facility Subcategory that identifies the point as either a Fuel Station or a Primary Health Centre.

   The Rural Facilities layer has more than 2,000 points for Primary Health Centres and Fuel Stations. The West Bengal Toll Plazas layer has 25 points for toll locations. You're looking for fuel stations that are 1,000 meters from a toll plaza, so you could manually search the map for each toll plaza point, measure 1,000 meters around it, and click each point within that range to find the fuel stations. But it's far more sensible to run a tool to do this task for you.

Search for locations using the Find by Attributes and Location tool

Next, you'll open and run the Find by Attributes and Location tool. This tool will search for features in the layer, based on values that are stored in the attribute table.

> *Review the basics for navigating the user interface in Map Viewer. The Contents toolbar is the dark vertical toolbar on the left, and the Settings toolbar is the light vertical toolbar on the right. When you click or select something on a toolbar, a pane usually opens, allowing you to adjust a related setting.*

1. On the Contents toolbar, click Layers.
2. In the Layers pane, click the Rural Facilities layer. On the Settings toolbar, click Analysis.
3. Click Tools.
4. Expand Find locations, and click Find by Attributes and Location.

   The tool pane opens.

Configure the Find by Attributes and Location tool

1. In the Find by Attributes and Location tool pane, click Build new query.

   The query builder opens, allowing you to define what you're looking for.

2. Under Find features from, click the down arrow and click Rural Facilities.

   The pop-up for the Rural Facilities layer showed a Facility Subcategory that listed two types of facilities—fuel stations and health centers. The query builder is now set to find features from the Rural Facilities layer and to narrow the features in that layer based on an attribute expression. You're on the right track!

   > **Hint:** The first step is to select only the Fuel Station points in the Rural Facilities layer. Later, you'll add a spatial expression to limit the search to points that are within 1,000 meters of toll plazas.

Query builder

Find features from

Rural Facilities

Attribute expression

Narrow down features of interest based on attribute values in a field.

Spatial expression

Narrow down features of interest based on spatial relationships with features in another layer.

Cancel Next

3. Confirm that the selections are correct and click Next.

   The query builder updates to ask which attributes from the Rural Facilities layer you're looking for. The objectid field is shown by default. To find the fuel stations, you'll narrow your search based on the Facility Subcategory attribute.

4. Click objectid, and click Facility Subcategory.

On your own

Read the query. Can you think of what you should enter in the last field (Enter value)? What kind of Facility Subcategory are you looking for?

5. Click the down arrow in the last field, and click Fuel Station so that your query now says this: Facility Subcategory equals Fuel Station.

Query builder

Find features from

Rural Facilities

Where

All of the following are true

Facility Subcategory equals Fuel Station

+ Attribute expression + Spatial expression Expression group

Cancel Add

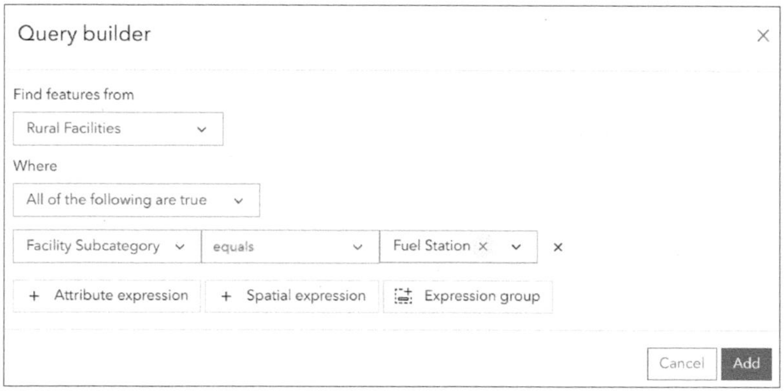

You've set the attribute part of your query to find the fuel stations. Now you'll add the location part.

6. Click Spatial expression.

   The default spatial operator is Intersects. The spatial operator describes how you'll search for location. You aren't looking for something that intersects with the fuel stations, so you'll change it.

7. Click Intersects and click Within a distance of.

   Now you'll specify the distance.

8. In the Enter value field, type **1000**. Make sure that the last field is set to Meters.

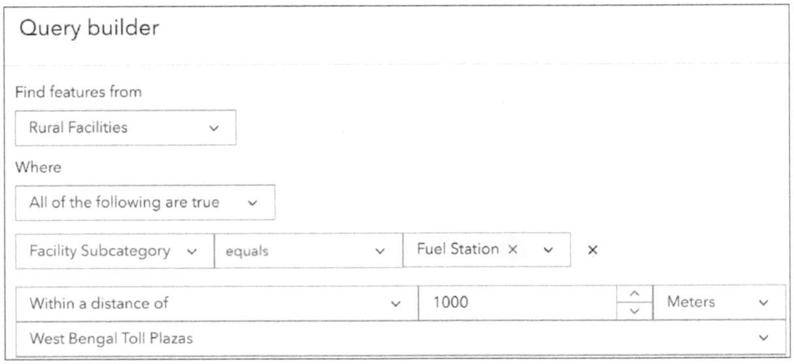

Query builder

Find features from

Rural Facilities

Where

All of the following are true

Facility Subcategory | equals | Fuel Station

Within a distance of | 1000 | Meters

West Bengal Toll Plazas

9. Read the final expression and click Add.

   Your query has been added to the tool. Next, you'll name the output layer and choose a location to store it.

10. For Output name, type **Fuel stations near toll plazas.**

11. For Save in folder, choose **Top 20 Tutorial Content.**

    > *If you need to create the folder, click the down arrow and choose Create new folder. Type* **Top 20 Tutorial Content.**

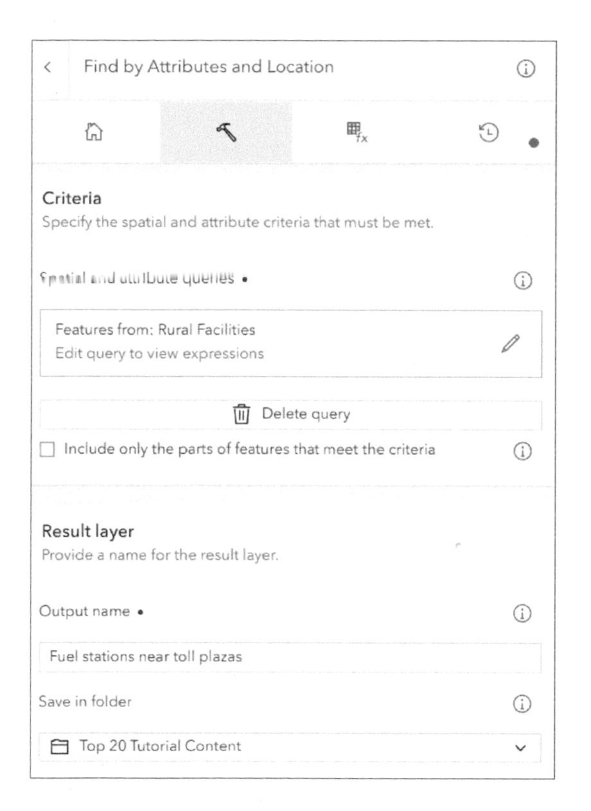

12. Click Estimate Credits.

13. Click Run, and wait for the layer to be added to the Layers pane.

> **Hint:** You can check a tool's progress by clicking History in the upper right of the tool pane. The History tab contains various history items. Later, if you want to change a setting and rerun the tool, you can click options and choose Open tool to see the settings you used, change them, and rerun. You can share this history with others who use your map.

Review your results

You'll review the new layer.

1. In the Layers pane, for the Rural Facilities layer, click the options button and click Remove.

   On your map, the blue points show fuel stations that are within 1,000 meters of toll plazas.

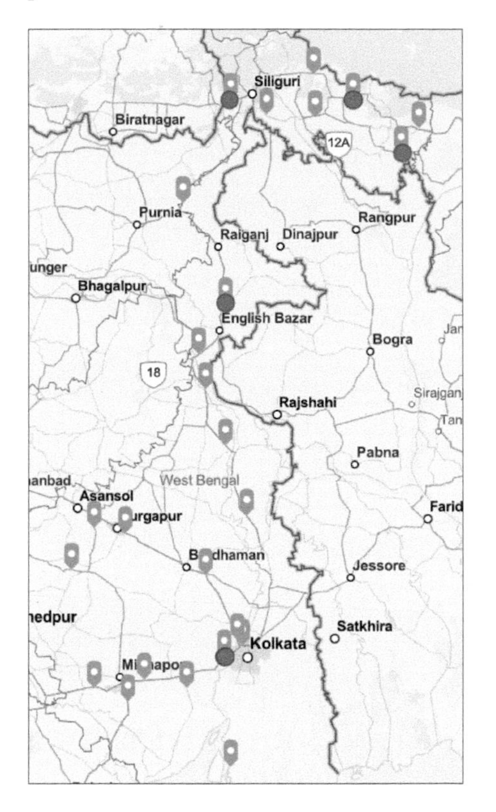

2. For the Fuel Stations Near Toll Plazas layer, click the options button and click Show table.

   Congratulations! Your search of West Bengal found nine fuel stations that met your criteria!

Fuel stations near toll plazas
9 records, 0 selected

On your own

Arrange and style the layers however you'd like.

Save your map

1. On the Contents toolbar, click Save and open > Save as.
2. For Title, type **Convenient Fuel Stations for Toll Plaza Stops <your initials>**.
3. For Folder, choose Top 20 Tutorial Content.
4. For Tags, type **Top 20, Fuel stations**.
5. For Summary, type **Fuel stations within 1,000 meters of toll plazas.**
6. Click Save.

Tutorial 8-2: Create and analyze walkable transit stops

You can create a feature based on the area that can be reached within a specified travel time. In the first tutorial, you searched for features, but now your analysis is focused on an area.

In this tutorial, you'll use the Generate Travel Areas tool to specify a travel time from a point. The result will be a polygon feature of the area that can be traveled within that time. You'll explore stations on a commuter rail system and create features of the areas that are within a 15-minute walk.

Open an existing map

First, you'll familiarize yourself with an existing map.

1. On the Contents toolbar, click Save and open > Open map. If you didn't complete tutorial 8-1, open a blank map.
2. Click My content and click ArcGIS Online.
3. Search for **owner:Top20EssentialSkillsForArcGISOnline transit** and press Enter.
4. On the Station Transit Areas map, click Open map.

The map contains a layer showing five stations on the Fitchburg commuter rail line outside the Boston metropolitan area.

Calculate walking travel areas

You want to find the distance you can walk from each station in 15 minutes and create polygon features for those areas.

1. On the Settings toolbar, click Analysis.
2. Click Tools.
3. Expand Use proximity and click Generate Travel Areas.
4. In the Generate Travel Areas tool pane, under Input layer, click Layer.
5. Click Commuter Stations.
6. Under Travel mode, click Driving Time and change the selection to Walking Time.

   > *The calculation will use 3.1 mph (5 kph).*

7. Under Cutoffs, type **15** and click Add.
8. Under Output name, type **Walkable station areas <your initials>**.
9. Under Save in folder, choose Top 20 Tutorial Content.
10. Click Estimate Credits.
11. Click Run, and wait for the layer to be added to the Layers pane.

    Once you configure the tool and click Run, your request is sent to the server for processing. A network service is used behind the scenes to calculate the distance you can travel.

> **Hint:** While you're waiting for the layer to be generated, click the information button (an i in a circle) at the top of the tool. Click Learn More to view the tool documentation web page. Whenever you want to learn more about a tool or setting, look here.

Review the travel areas

The features in the Walkable Station Areas layer show areas that can be reached in a 15-minute walk from the stations.

1. In the Layers pane, click the Commuter Stations layer and move it to the top of the layer list.

   Layers

   Commuter Stations

   Walkable station areas

   Setting the layer hierarchy so that Commuter Stations is drawn on top of Walkable Station Areas makes the map easier to understand.

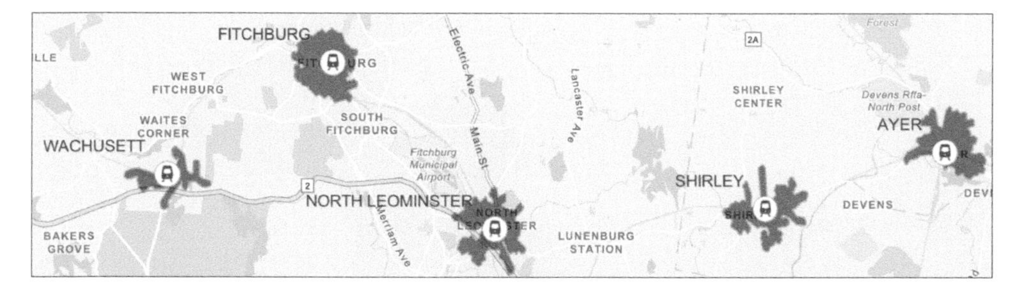

2. Click several of the walkable station area polygons. In the pop-ups, find the Area field. Compare the walkable areas.

   The walkable area around each station can vary significantly, determining how convenient each station is.

Save the map

1. On the Contents toolbar, click Save and open > Save as.
2. Save your map with the following settings:
   - Title: **Train station walkability**
   - Folder: Top 20 Tutorial Content
   - Tags: **Top 20, ArcGIS Online, walkability**
   - Summary: **Station walkability on the Fitchburg line, Boston, MA.**

3. On the Contents toolbar, click Share map.
4. In the Share pane, click Everyone (public) and click Save.

Take the next step

Adding a layer with census data will give you a deeper perspective on how high-value land near transit stations is being used. Seeing the population density in and around these areas can give you a good idea of how many people can easily walk to stations.

1. Add a layer from ArcGIS Online by searching for **owner:Top20EssentialSkillsForArcGISOnline Census.**
2. Add the Commuter Station Census Blocks layer to the map.
3. Move the layer to the bottom of the Layers pane.

   Drawing the layers in this order gives you a visual sense of the population density around the transit stops. Next, you'll estimate how many people live within a 15-minute walk of the stations.

4. In the Tools list, expand the Summarize data group. Configure the Summarize Within tool using the following parameters:
5. For Input features, choose Commuter Station Census Blocks.
6. For Summary polygon layer, choose Walkable station areas.
7. Turn off Summarized geometry of features.
8. For Field statistics, choose Total Population.
9. For Output name, type **Population living near transit stations.**
10. Click Estimate Credits, and run the tool.
11. Turn off visibility for Walkable station areas and Commuter Station Census Blocks.
12. Click a Population living near transit stations feature. In the pop-up, read the Sum Total Population field.

    This field indicates the number of people estimated to live within the walkable area.

13. Compare the Sum Total Population for the Fitchburg and Wachusett stations.

    > *This total population summary is an estimate. If 50 percent of a census polygon with 100 people is overlapped by the walkability polygon, 50 people are estimated to live in the walkability polygon.*

    This layer gives a good estimate of how many people can easily walk to the station.

14. Change the basemap to Imagery to see how the land around the stations is used.

15. Save your map.

Summary

In this chapter, you created a query to locate things based on what and where they are. You also defined an area based on travel time, creating features that indicated where someone can travel in a given amount of time. These are powerful search methods for locating things and places.

Workflows

Find by Attributes and Location tool

1. Build the attribute query, selecting the layer and attributes of interest.
2. Add a spatial expression with a spatial operator that defines the location.

Generate Travel Areas tool

1. Select the input layer.
2. Determine the travel mode.
3. Enter cutoff times and units.
4. Choose a travel direction.
5. Choose a departure time.
6. Choose an overlap policy.

CHAPTER 9
Creating and analyzing imagery

Objectives

- Learn about tiled and dynamic imagery layers.
- Create a tiled imagery layer and add it to your contents.
- Perform a burn analysis using a raster function template.
- Reclassify and create a final tiled imagery layer.

Introduction

In this chapter, you'll upload an image to your account and perform analysis. To start, you'll download an image taken from the Landsat 8 satellite and upload it to your contents as a tiled imagery layer.

A hosted imagery layer is a type of imagery layer that can be made available in a browser but still has the pixel values that were recorded when the image was taken. This feature allows you to access those values and use them for analysis or incorporate the layer in a web map or app.

You can publish hosted imagery layers as tiled imagery layers or dynamic imagery layers:

- Tiled imagery layers are static tiles hosted in ArcGIS Online that are sent to the browser, which decodes the imagery tiles, extracts bands, and renders the imagery.
- Dynamic imagery layers are a single image or collection processed as needed by the server, based on the required projection, bands, or other processing. The result is sent to the browser.

If you want to add an imagery layer to your map and share the result publicly, a tiled imagery layer is the best option. You'll explore this workflow in the following tutorial.

Tutorial 9-1: Upload a satellite image and create a tiled imagery layer

In this tutorial, you'll download an image taken by the Landsat 8 satellite. The image shows an area in Australia that was affected by wildfire. You'll upload the image and create a hosted tiled imagery layer. Next, you'll analyze the burned area using a raster function template. The resulting image showing wildfire impacts will be available for analysis or sharing in an app.

Sign in and download imagery files

You'll sign in and download a zip file containing the Landsat 8 imagery.

1. In a browser, sign in to ArcGIS Online.
2. On the navigation bar, click the magnifying glass to search for **owner:Top20EssentialSkillsForArcGISOnline wildfire** and press Enter.
3. Under Filters, turn off the filter for your organization and expand your search to all of ArcGIS Online.
4. On the Australia Wildfire item, click Download. Download the file to a convenient location on your computer.

> *This file size is about 1 gigabyte and will take some time to download. You'll be uploading the files to this folder and won't need the files after that, so don't worry about storing this file in a permanent location. You can delete it after you've finished this chapter.*

5. Unzip the file.

   A folder is created containing a Landsat 8 scene made up of a series of files.

Create a tiled imagery layer

You'll upload the imagery and configure the properties.

1. On the ArcGIS Online navigation bar, click Content.
2. Click New item.
3. In the New item window, click Imagery layer.

4. In the Create imagery layers window, confirm that Tiled Imagery Layer is selected and click Next.

   Next, you'll specify multiple Landsat files to create a mosaicked image.

5. Click One Mosaicked Image and click Next.

   Now you'll set the raster type.

6. Click Raster Dataset. From the list, click Landsat 8.
7. Click Configure properties.
8. In the Raster type properties window, confirm that the Processing tab is selected. Under Processing templates, click Pansharpen to view the list.

   This list shows available processing templates, based on the raster type you've selected.

9. In the list, click Multispectral.

Raster type properties
General Processing Spatial Reference Metadata
Processing templates
Multispectral

10. Click Apply.

    The input Landsat image contains 11 raster bands. When you choose the Multispectral processing template, only eight will be used.

11. In the Create imagery layers window, under Select input imagery, click Browse, browse to the folder you created, and select it.

    > *You can also drop the folder you plan to upload.*

    The process of uploading the files will begin.

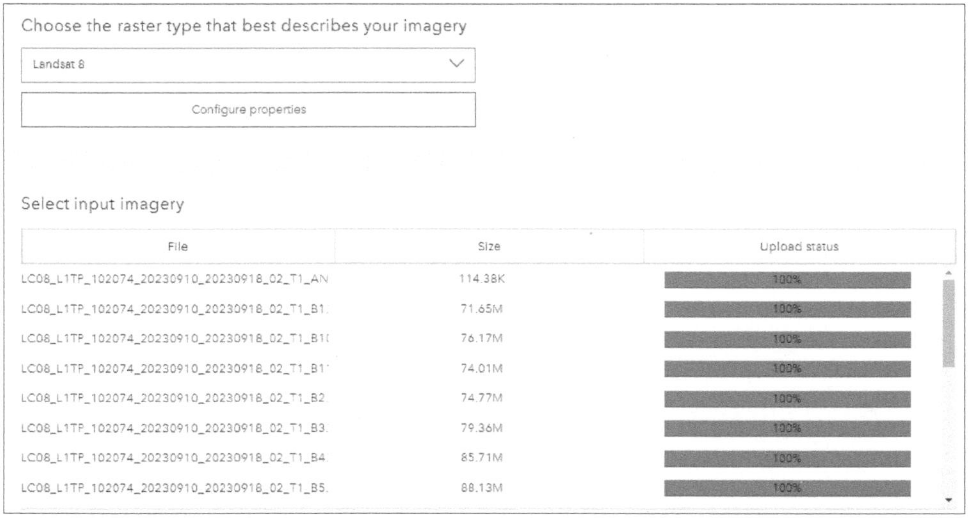

Choose the raster type that best describes your imagery

Landsat 8

Configure properties

Select input imagery

File	Size	Upload status
LC08_L1TP_102074_20230910_20230918_02_T1_AN	114.38K	100%
LC08_L1TP_102074_20230910_20230918_02_T1_B1	71.65M	100%
LC08_L1TP_102074_20230910_20230918_02_T1_B1(	76.17M	100%
LC08_L1TP_102074_20230910_20230918_02_T1_B1'	74.01M	100%
LC08_L1TP_102074_20230910_20230918_02_T1_B2	74.77M	100%
LC08_L1TP_102074_20230910_20230918_02_T1_B3	79.36M	100%
LC08_L1TP_102074_20230910_20230918_02_T1_B4	85.71M	100%
LC08_L1TP_102074_20230910_20230918_02_T1_B5	88.13M	100%

12. Click Next.

    Now you'll name the layer.

13. Save your item with the following settings:

    - Title: **Satellite image <your initials>**
    - Tags: **Top 20, satellite**
    - Summary: **A satellite image for use in wildfire impact analysis.**
    - Folder: Top 20 Tutorial Content

    > *If you haven't created the Top 20 Tutorial Content folder in an earlier tutorial, you can save your item in your root folder.*

14. Click Create.

Hint: You can use Notebooks to move images from a provider and add them directly to your contents in ArcGIS Online, where they can be made available as tiled imagery layers without the need to download and then upload. That advanced workflow isn't covered here, but when you're ready, you can visit links.esri.com/ImageryDirect to learn more.

After the uploading and processing are complete, a tiled imagery layer is created and you'll be directed to the item page.

Analyze the image in Map Viewer

Next, you'll open the tiled imagery layer in Map Viewer and identify the burned areas with raster analysis tools.

1. Review the item details and click Open in Map Viewer.

   Map Viewer opens with the image correctly positioned over the basemap.

2. On the Settings toolbar, click Styles.
3. Under RGB, click Style options.
4. In the Style options pane, under Apply dynamic range adjustment, click the toggle button.

   This adjustment improves the image display, making the visualization clearer.

5. On the Settings toolbar, click Analysis.

> *Review the basics for navigating the user interface in Map Viewer. The Contents toolbar is the dark vertical toolbar on the left, and the Settings toolbar is the light vertical toolbar on the right. When you click or select something on a toolbar, a pane usually opens, allowing you to adjust a related setting.*

6. In the Analysis pane, click Raster Functions.

   The list of available raster functions appears.

7. At the bottom of the Raster Functions pane, click the arrow next to New Raster Function Template and choose Open Raster Function Template.

   The Browse raster function templates window appears.

   Raster function templates are combinations of raster functions that are used to create a specific visualization or analysis. Processing steps can be chained together to perform a customized workflow. You can select any raster function that's been shared with you or that can be found on ArcGIS Online or ArcGIS Living Atlas.

8. In the Browse raster function templates window, click My content and click ArcGIS Online.

9. In the search box, type **Normalized Burn Ratio**.

10. On the Normalized Burn Ratio (NBR) Analytic item, click More details.

11. Review the details in the Overview section to learn more.

    This raster function template will correctly perform the desired calculation using the supplied imagery.

12. Click the Normalized Burn Ratio (NBR) Analytic item and click Confirm.

Configure the Normalized Burn Ratio analysis

You'll configure the Normalized Burn Ratio tool.

1. In the Normalized Burn Ratio (NBR) Analytic tool pane, under Raster, click Select Layer. From the list, click Satellite image.

2. For Output name, type **BurnAnalysis <Your Initials>**.

3. For Result type, confirm that Tiled imagery layer is already selected.

This is the correct selection. Similar to the layer that's used in the map, the output will also be a tiled imagery layer.

4. Under Save in folder, choose Top 20 Tutorial Content.
5. Click Estimate credits.
6. Click Run.

> *To view the tool's progress, click History at the top of the pane.*

When processing completes, the BurnAnalysis layer is added to the map. The layer shows the burned areas as black.

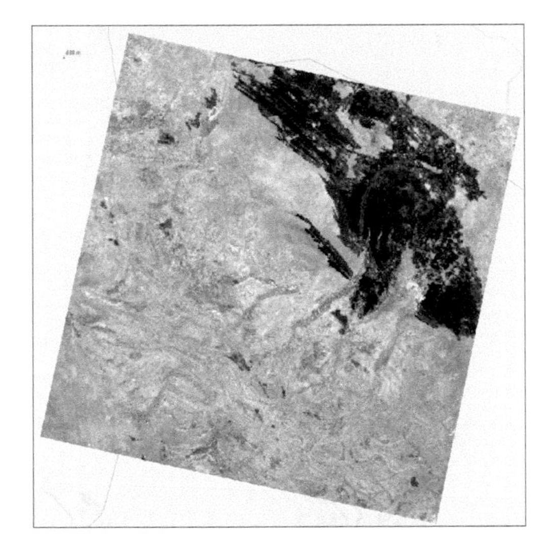

Identify and classify burned areas

Next, you'll identify the pixel values that indicate burned areas and style those areas.

1. On the map, click the image to view the pop-up.

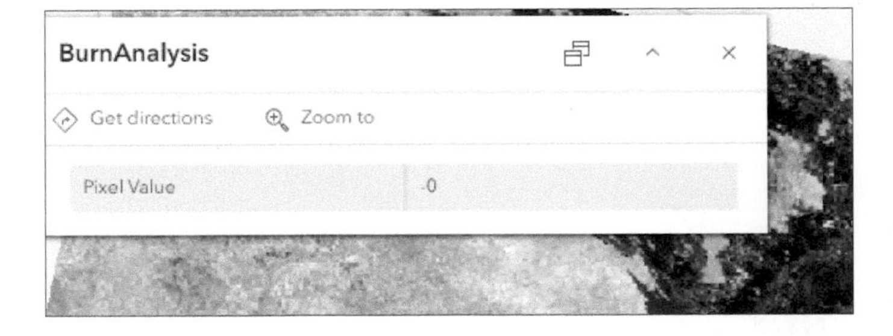

The pop-up shows a rounded number. You'll adjust the settings to see the full pixel value.

2. On the Settings pane, click Fields.
3. In the Fields pane, click Pixel Value.
4. Under Significant digits, click the down arrow and choose 8 Decimal places.

Formatting ×
Display name
Pixel Value
Significant digits
8 Decimal places
Show thousands separator
1 selected ×
Pixel Value {Raster.ServicePixelValue} 123
Raw Pixel Value {Raster.ServicePixelValue.Raw} 123

5. Click Done.

   The pixel value in the pop-up updates to show the full value, based on the location you clicked. You can now view the pixel values of any location in the image.

6. With the BurnAnalysis layer selected in the Layers pane, on the Settings toolbar, click Styles.
7. Click Classify.

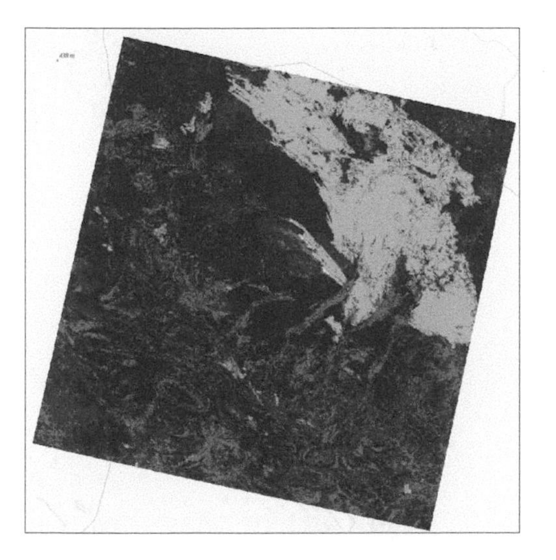

8. Under Classify, click Style options.

   As shown in the pane, the default method, natural breaks, is applied. This method creates five classes, with the burned area shown in yellow and light orange. On the map, the legend shows the pixel value ranges for each class. An estimate of pixel values for burned locations would include the pixel values from –0.04 through –0.769.

BurnAnalysis
0.125 - 0.984
0.07 - 0.125
-0.04 - 0.07
-0.32 - -0.04
-0.769 - -0.32

Create a layer of burned areas

Now that you've identified pixel values for areas that burned, you'll create a layer of only those areas.

1. On the Settings toolbar, click Analysis.
2. At the top, click the Raster Functions button.

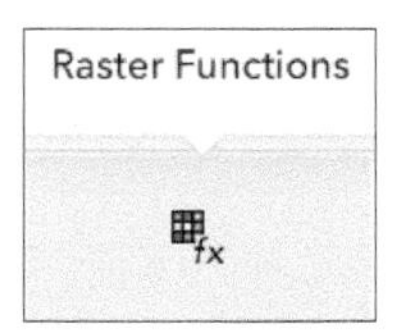

3. In the search box, type **Remap**.

Raster Functions
Remap
Remap

4. In the results, click Remap.
5. Under Raster, click Select Layer and click BurnAnalysis.
6. For Minimum, type **-0.769**.
7. For Maximum, type **-0.04**.
8. For Output, type **1**.
9. Check the Change missing values to NoData check box.

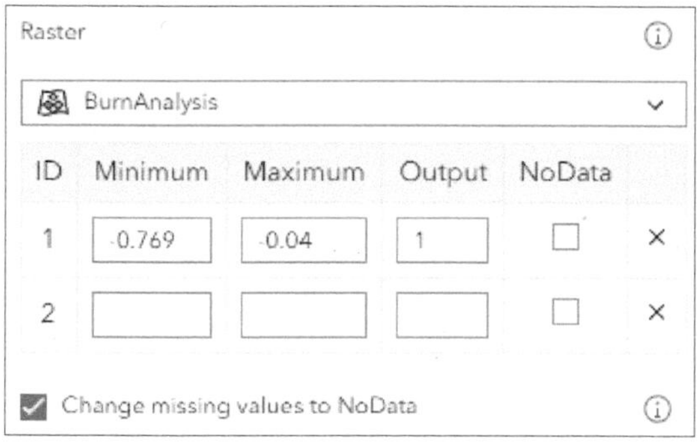

Raster

BurnAnalysis

ID	Minimum	Maximum	Output	NoData	
1	-0.769	-0.04	1	☐	×
2				☐	×

☑ Change missing values to NoData

These settings will change the value of any pixels that fall in the specified range to a value of 1. All other values will be changed to NoData.

10. Under Show Preview, click the toggle button.

This tool takes processing time—it adds a layer to your contents. Before you run the tool, you should preview it to make sure it will generate the expected result. In this case, you should expect a layer of only burned areas to be shown, since the tool is set to remap burned locations to have a pixel value of 1 and all other pixels to have NoData.

11. On the map, confirm that your preview displays burned areas in black.
12. Under Output name, type **BurnedAreas <Your Initials>**.
13. Under Result type, confirm that you're creating a tiled imagery layer.
14. Under Save in folder, choose Top 20 Tutorial Content.
15. Click Estimate Credits.
16. Click Run.

The BurnedAreas layer is added to the map.

Visualize the burned areas on the map

Now you'll visualize only the burned areas.

1. On the Contents toolbar, in the Layers pane, point to the BurnAnalysis layer and click the visibility button (eye) to turn off the layer's visibility.

   The BurnedAreas layer shows burned areas in black.

   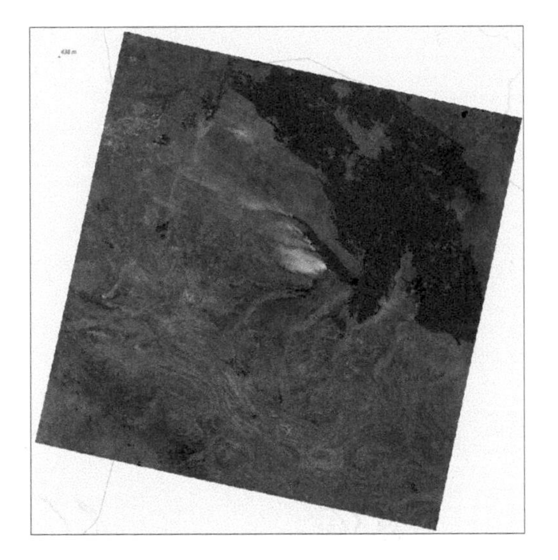

   The burned areas are difficult to identify on the satellite imagery, so you'll restyle the BurnedAreas layer.

2. In the Layers pane, select the BurnedAreas layer. On the Settings toolbar, click Styles.
3. Under the Stretch style, click Style options.
4. Under Color scheme, click the default color ramp.
5. In the Color scheme pane, click the default color ramp, and click the Yellow to Red color ramp.

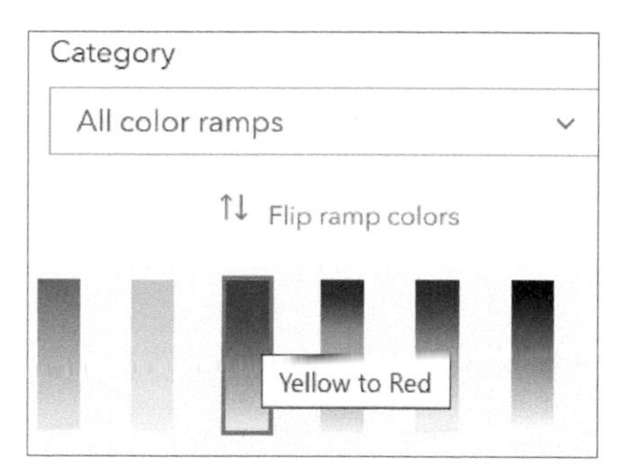

The layer style is updated on the map to make the burned areas much easier to see.

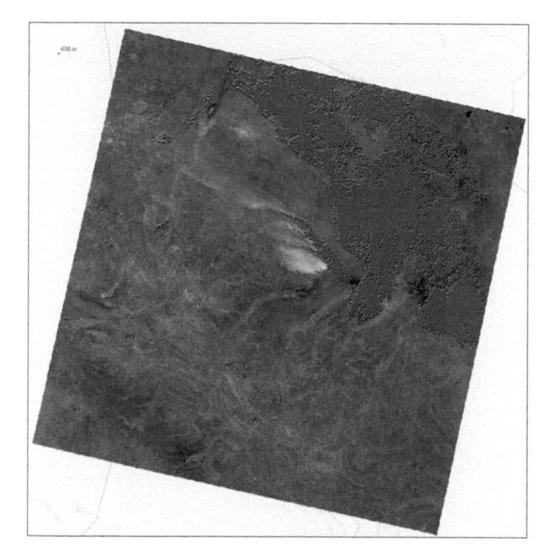

6. Click Done, and close any open panes.

   The new visualization makes the burned areas easier to analyze later.

Save your map

1. On the Contents toolbar, click Save and open > Save as.
2. Save your map with the following settings:
   - Title: **Burn imagery analysis**
   - Folder: Top 20 Tutorial Content
   - Tags: **Top 20, tiled imagery**
   - Summary: **A map showing burned areas.**

3. Click Save.

4. On the Contents toolbar, click Share map.

5. In the Share pane, click Everyone (public) and click Save. Share all files in your map.

Take the next step

A huge catalog is available of public-domain satellite imagery from USGS EarthExplorer. You can use this imagery to perform similar analyses for any wildfire. Browse the catalog at earthexplorer.usgs.gov. Specify the fire location on the map and choose a date range after the fire, giving yourself a wide margin to find a time when a satellite image may have been taken. Specify a dataset of Landsat, and choose the Landsat Collection 2 Level-1. In the search results, preview potential satellite imagery by clicking Show Browse Overlay to confirm that the image is suitable. Download all the files for the selected scene and repeat this tutorial for your location using that data.

Summary

In this chapter, you created a tiled imagery layer and performed a raster analysis to determine locations that had burned. You learned how to view specific pixel values in the imagery layer. Finally, you created a tiled imagery layer showing only burned areas.

Workflow

1. Create a new item.
2. Choose Imagery layer.
3. Choose Tiled Imagery layer.
4. Choose a layer configuration.
5. Choose the raster type and properties.
6. Add and upload the imagery file or files.
7. Add the item details.

CHAPTER 10

Configuring charts

Objectives

- Select appropriate chart types based on your data.
- Create ArcGIS Online Map Viewer charts to explore data and communicate results.
- Join a table to a feature layer.
- Customize chart elements, including titles, axes, colors, and legends.
- Share maps and charts in a web app using Chart Viewer in ArcGIS Instant Apps (optional).

Introduction

In this chapter, you'll learn how to use charts in ArcGIS Online Map Viewer to perform exploratory data analysis. Like maps, charts are a common way to visualize and explore data. By using graphical representations, charts make it easier to visualize distributions and uncover trends, outliers, or patterns in the data. In addition, charts help you interpret analysis results and communicate findings. Different types of charts are supported in Map Viewer, including bar charts, line charts, pie charts, histograms, scatterplots, and box plots. The best chart type depends on the data you're trying to visualize.

In this chapter, you'll use maps and charts to explore electric car adoption patterns in the United States and identify which states have the highest adoption of electric vehicles.

Tutorial 10-1: Chart the table of a vehicle registration count

In this tutorial, you'll create a bar chart to compare the count of electric vehicles by state with the vehicle registration count data made available by the US Department of Energy.

Sign in and download a CSV file

You'll start by signing in to ArcGIS Online and downloading a CSV file of vehicle registration counts by state.

1. Sign in to ArcGIS Online.

> **Hint:** If needed, review the detailed steps for signing in at the beginning of chapter 1.

2. On the top navigation bar, click the magnifying glass to search for items in ArcGIS Online.
3. Type **owner:Top20EssentialSkillsForArcGISOnline vehicle** and press Enter.
4. Under Filters, turn off the filter for your organization and expand your search to all of ArcGIS Online.
5. On the Vehicle Registration Counts By State item, click Download. Save the file to a convenient location on your computer.

   The VehicleRegistrationCountsByState.csv file contains data on vehicle registration counts by US states. The file has counts for different vehicles, including traditional gasoline cars and cars using renewable energy such as electricity, biodiesel, hybrid, and so on.

Open a new, blank map and add a table using a CSV file

Now you'll open a new web map and add the VehicleRegistrationCountsByState.csv file to your map as a table.

1. On the top navigation bar, click Map.

> *Review the basics for navigating the user interface in Map Viewer. The Contents toolbar is the dark vertical toolbar on the left, and the Settings toolbar is the light vertical toolbar on the right. When you click or select something on a toolbar, a pane usually opens, allowing you to adjust a related setting.*

2. On the Contents toolbar, in the Layers pane, click Add > Add layer from file.
3. In the Add Layer window, click Your device and browse to the VehicleRegistrationCountsByState.csv zip file, or drag it into the window.
4. Confirm that Create a hosted feature layer and add it to the map is selected, and click Next.

   The 13 listed fields match the column headers in the CSV file.

5. Click Next.

   The Location settings step gives you the option to define locations for the data using addresses or place-names. The file doesn't contain location data, so you'll add the data as a nonspatial table layer instead.

6. For Location settings, at the top, click Addresses or place names and choose None.

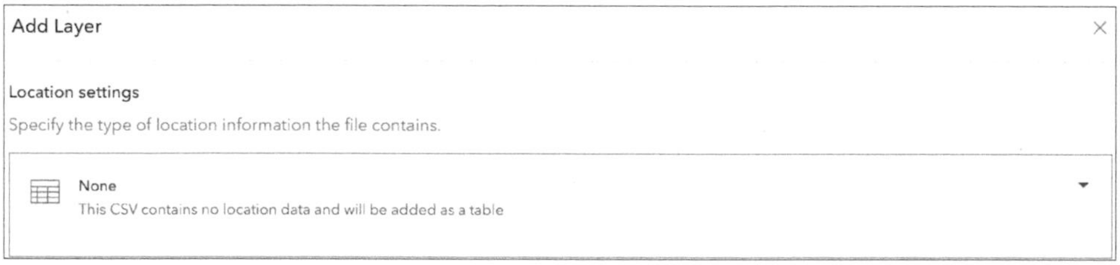

Add Layer
Location settings
Specify the type of location information the file contains.
None
This CSV contains no location data and will be added as a table

7. Click Next.

8. Add the following settings:

   - Title: **Vehicle Registration Counts By State <your initials>**
   - Folder: Top 20 Tutorial Content
   - Tags: **Top 20, ArcGIS Online, Vehicle**
   - Summary: **Vehicle Registration Counts By State, 2022.**

   > *If you need to create the folder, click the down arrow and choose Create new folder. Type* **Top 20 Tutorial Content**.

9. Click Create and add to map.

10. On the Contents toolbar, click Tables.

    The Vehicle Registration Counts By State table is listed in the Tables pane.

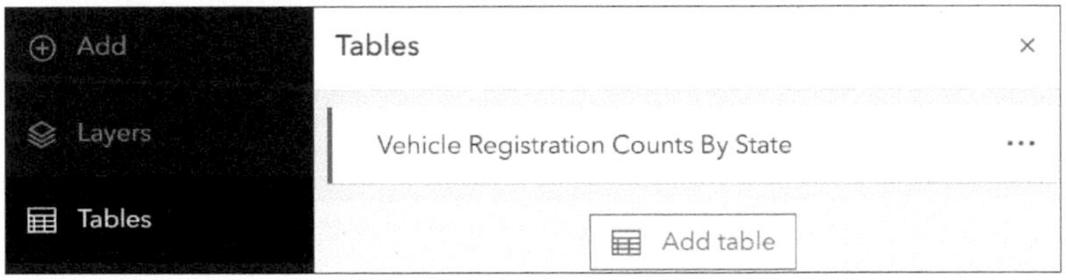

Examine the attributes

Next, you'll open the table to review the attributes and ensure that the data was imported correctly.

1. In the Tables pane, click Vehicle Registration Counts By State to open the table.

2. Review the table's contents to confirm that the data was imported correctly.

Vehicle Registration Counts By State

State	Electric (EV)	Plug-In Hybrid Ele.
Alabama	8,700	4,400
Alaska	2,000	700

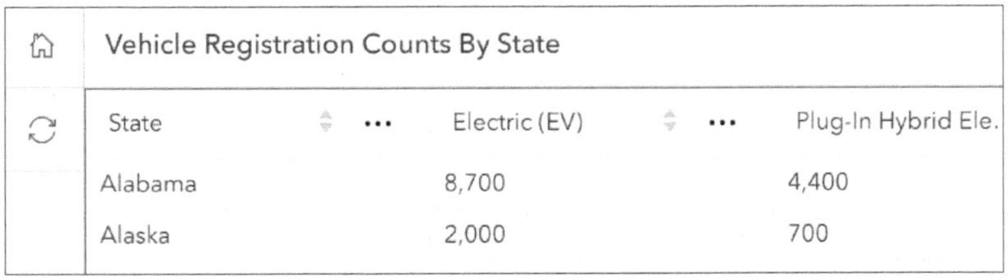

3. Close the table.

Chart the table

Next, you'll create a bar chart to compare the counts of electric vehicles by state. Bar charts show comparisons of values in different categorical variables, such as race, age groups, classes, and so on.

1. On the Settings toolbar, click Configure charts. In the Charts pane, click Add chart and click Bar chart.

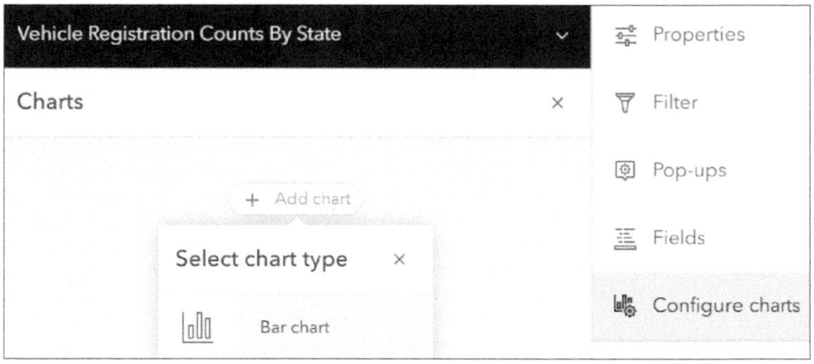

2. On the Data tab, for Category, choose State.

3. Click Select numeric fields.

4. Choose Electric (EV) and click Done.

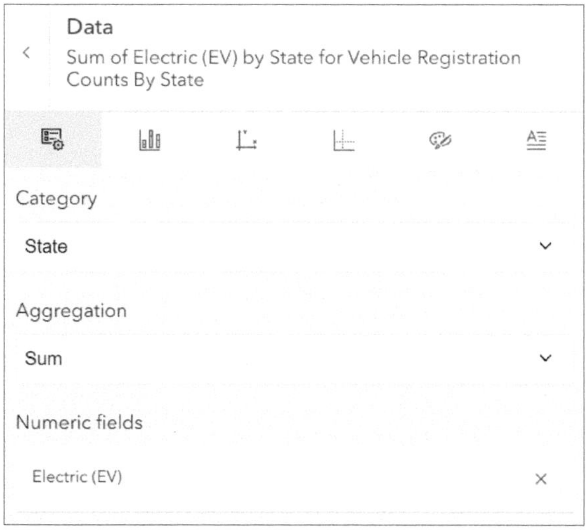

A bar chart showing the count of electric vehicles appears.

5. Under Sort order, click the down arrow and choose Y-axis descending.

6. At the top of the pane, click the General tab.

7. For Chart title, delete the default title and type **Number of electric vehicles by state.**

8. For Y-axis title, delete the default title and type **Number of electric vehicles.**

9. On the chart, point to some of the bars, including the top three highest bars.

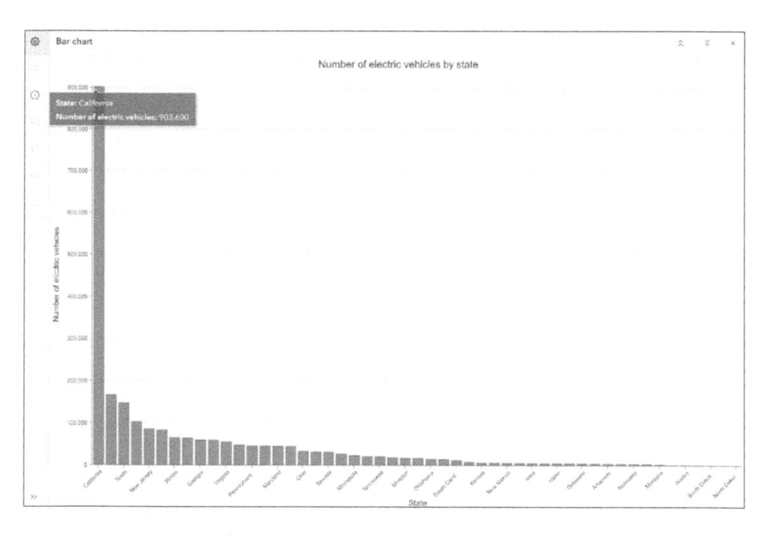

You can see that California, Florida, and Texas have the highest number of electric cars.

Hint: To see the charts you've made, in the Contents pane, click Charts.

Save the web map

Next, you'll save the web map and assign it a title, tags, and a summary to make it easy to find and identify later.

1. On the Contents toolbar, click Save and open > Save as.
2. In the Save map window, add the following settings:
   - Title: **Electric Vehicles by State <your initials>**
   - Folder: Top 20 Tutorial Content
   - Tags: **Top 20, ArcGIS Online**
   - Summary: **A map showing electric vehicles by state.**
3. Click Save.

Tutorial 10-2: Map and chart the count of electric vehicles by state, normalized by population

In the previous tutorial, you created a bar chart based on the VehicleRegistrationCountsByState.csv table. The chart shows that California has the highest number of electric cars, followed by Florida and Texas. However, California is also the most populous state in the United States; similarly, Texas and Florida also have high populations and high numbers of electric cars. A more meaningful comparison would be to normalize the electric vehicle counts by population.

Next, you'll use the Join Features analysis tool to join the table to a US state boundaries layer in ArcGIS Living Atlas. The result from the Join Features tool will be a feature layer with the additional population attribute for each state. You'll be able to visualize the data on the map and configure interactive charts to analyze the adoption patterns of electric cars across the United States.

Join the vehicle registration counts table to the USA State Boundaries layer

1. If you closed it earlier, open the web map you created in tutorial 10-1.
2. On the Settings toolbar, click Analysis.
3. In the Analysis pane, click Tools.
4. Expand Summarize data and click Join Features.

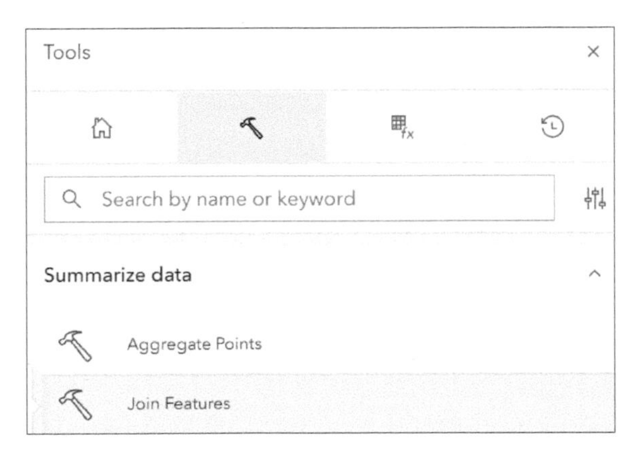

5. In the Join Features pane, under Target layer, click Layer.

> **Hint:** To learn what a target layer is, in the pane, at the end of the Target layer input, click the information button (a lowercase i) to read more.

You'll browse ArcGIS Living Atlas to find the target layer you'll use.

6. In the Select layer pane, at the bottom, click Browse layers.
7. Click My content. In the list, click Living Atlas analysis layers.
8. In the search bar, type **USA States**.
9. In the search results, on the USA States Generalized Boundaries item, click Select layer and choose states.
10. At the bottom of the window, click Confirm.

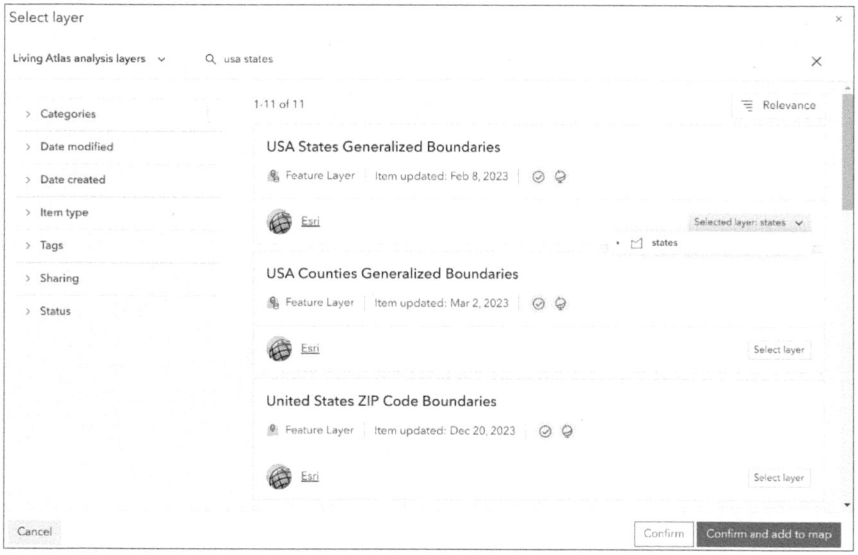

11. In the Join Features tool pane, under Join layer, click Layer and choose Vehicle Registration Counts By State.
12. Under Join settings, set or verify the following parameters:
    - Use attribute relationship: enabled
    - Target field: State Name
    - Join field: State

- Join operation: Join one to one
- Multiple matching records: Only keep first matching record
- Join type: Left join

> **Hint:** If you aren't familiar with these join settings, click information next to a setting to learn more.

13. For Output name, type **Vehicle Registration Counts By State <your initials>**.

14. For folder, choose Top 20 Tutorial Content.

15. Click Estimate Credits.

16. Click Run.

    When the tool completes, the output appears on the map.

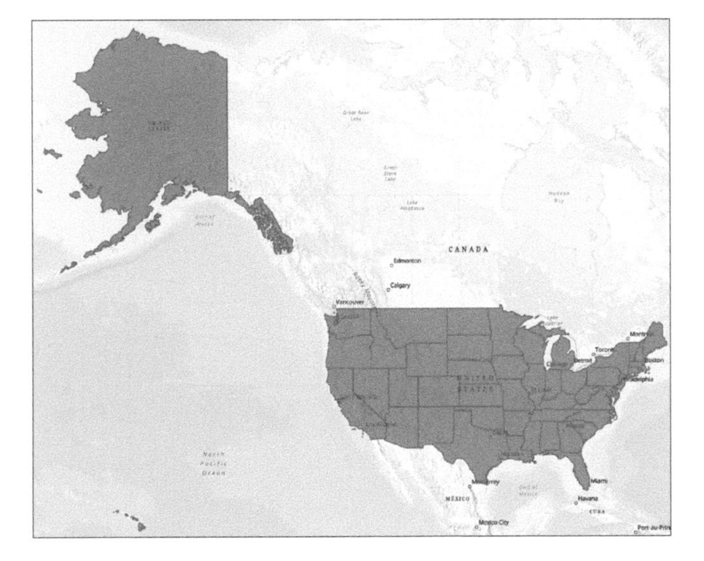

Symbolize layers

Next, you'll style the data to show the counts of electric cars in each state, normalized by population on the web map.

1. On the Contents toolbar, in the Layers pane, click the Vehicle Registration Counts by State layer to select it.

2. On the Settings toolbar, click Styles.

3. In the Styles pane, under Choose attributes, click Field.
4. In the search box, type **El**, and choose Electric (EV).

> **Hint:** Typing part of a search term to narrow a result list lets you quickly find what you're looking for.

5. Click Add.
6. In the Styles pane, under Pick a style, click Counts and Amounts (color).
7. Under Counts and Amounts (color), click Style options.

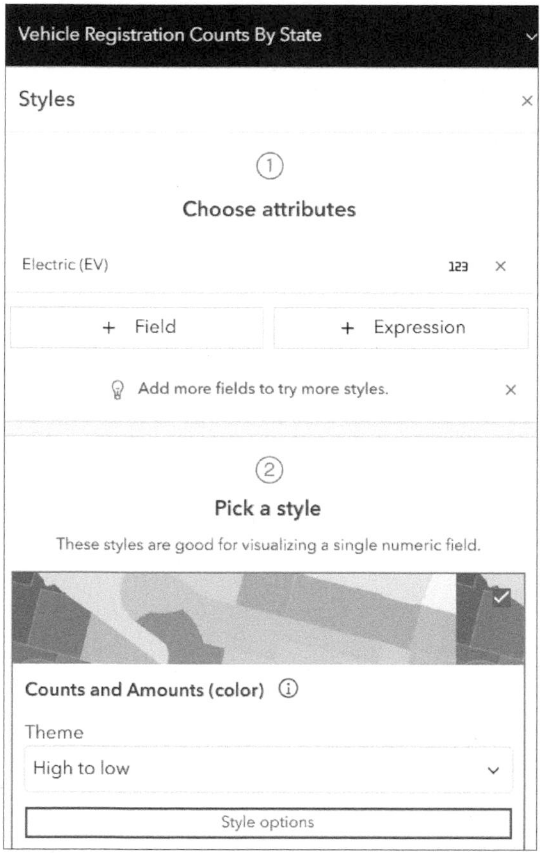

8. In the Style options pane, under Divided by, click the down arrow and choose POPULATION_2020.

   The completed style settings are shown.

Style options
Vehicle Registration Counts By State

Counts and Amounts (color)

Electric (EV)

Theme

High to low
Vary the color of features from high to low.

Divided by

POPULATION_2020

9. Click Done and click Done.

10. On the Contents toolbar, click Legend.

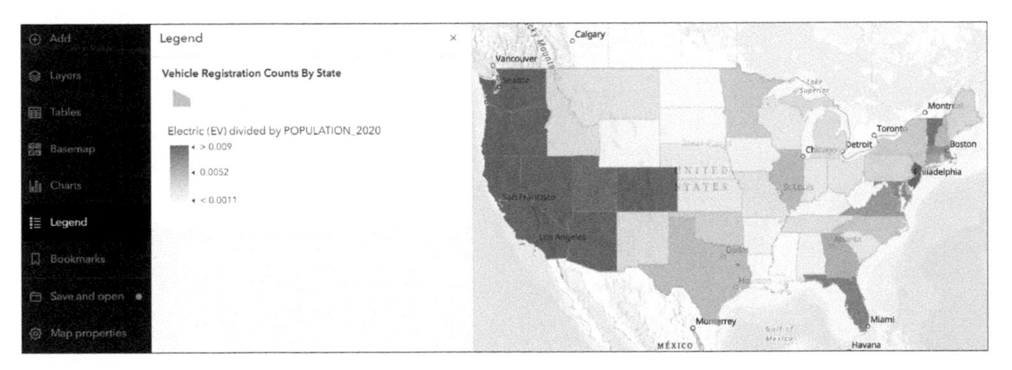

The map now shows the number of electric vehicles in each state normalized by population, with darker colors representing higher numbers of electric vehicles per capita and lighter colors representing lower numbers. Although California still has a large number of EVs per capita, Florida and Texas are no longer in the top three.

11. On the Contents pane, click Save and open > Save.

Explore relationships using scatterplots

Here you'll create a scatterplot to explore the correlation between the state's population and registered electric cars. Scatterplots are used to measure the strength of a relationship between two numeric variables.

> **Hint:** The type of chart you use depends on the type of data in the layer and the information you want to present. Bar charts summarize and compare categorical data, histograms show the distribution of a continuous numeric variable, and scatterplots visualize relationships between numeric variables.

1. On the Content pane, click Layers.
2. In the Layers pane, select the Vehicle Registration Counts By State layer.
3. On the Settings toolbar, click Configure charts.
4. In the Charts pane, click Add chart and choose Scatterplot.
5. On the Data tab, under Variables, for X-axis number, choose POPULATION_2020. For Y-axis number, choose Electric (EV).
6. Under Statistics, turn on Show linear trend.

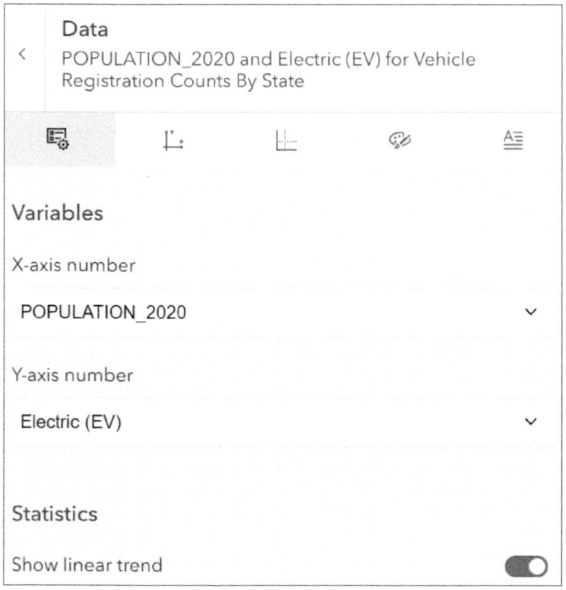

You now have a scatterplot with a line of the best fit and a correlation efficient (R-squared) value of 0.62, indicating a positive linear relationship.

> **Hint:** R-squared value measures the strength of the relationship between two numeric variables. It's a value between 0 and 1. The higher the R-squared value, the stronger the linear relationship is for the variables.

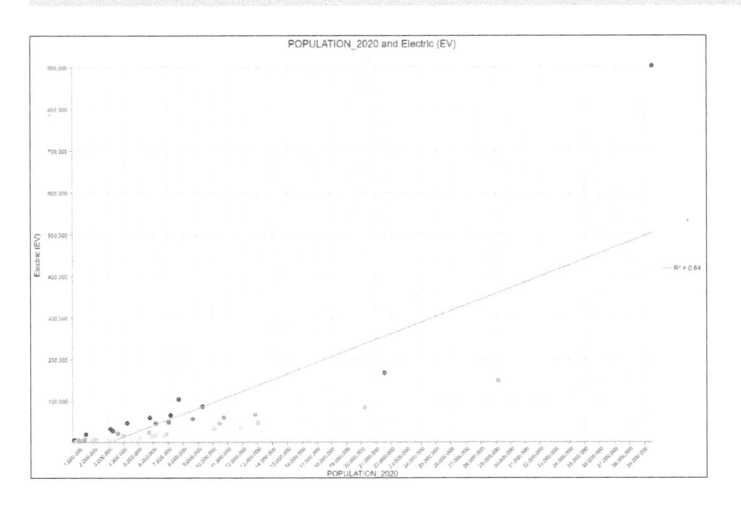

By default, the scatterplot uses the same colors as the source layer.

7. At the upper right of the chart, click Collapse.
8. Position the map and chart so that you can see both at the same time.

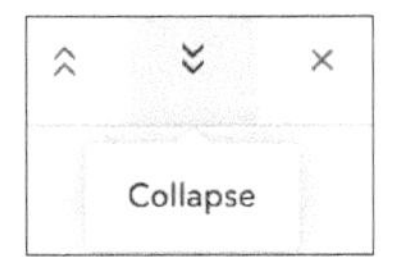

The scatterplot shows a general linear relationship between the number of electric vehicles and the state's population. In other words, the more people in the state, the higher the count of electric cars registered. However, you can also observe some data points that don't fit the pattern (points that are well above or below the line of the best fit). These data points are called outliers.

9. In the upper right of the scatterplot, click the data point that's well above the line of the best fit.

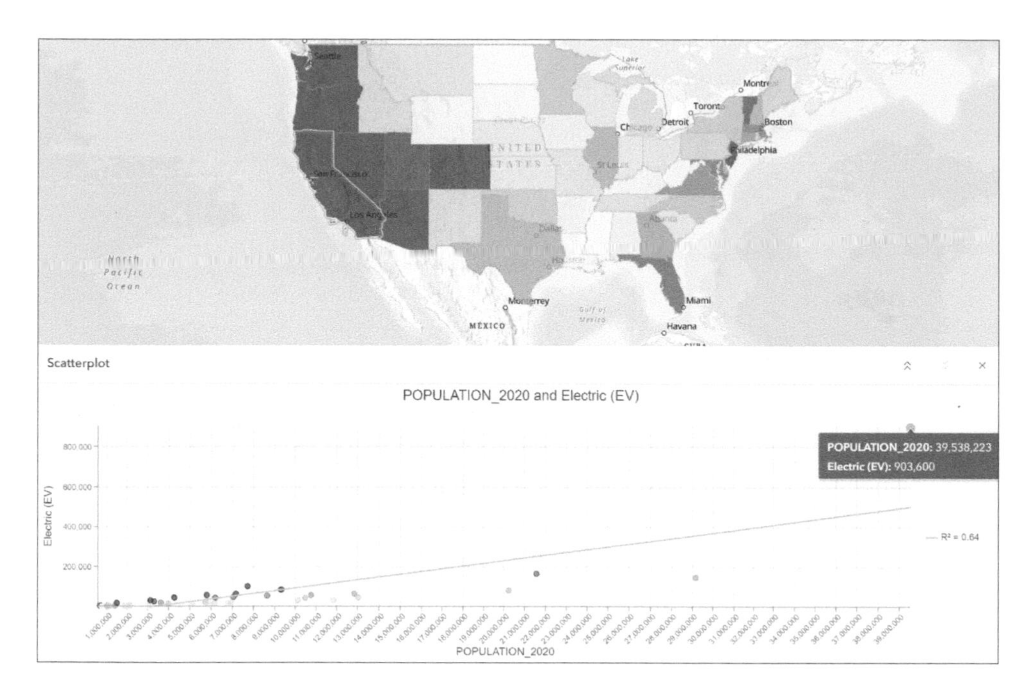

The corresponding state (California) is highlighted on the map. In this case, the registered number of electric vehicles in California, based on its population, is significantly higher than the predicted value (which would be on the line).

10. In the lower right of the scatterplot, click the data point that's well below the line of the best fit.

    Now the corresponding state (Texas) is highlighted on the map. In this case, the registered number of electric vehicles in Texas, based on its population, is significantly lower than the predicted value (which would be on the line).

On your own

On the scatterplot, select different points above and below the line of the best fit, and look at the corresponding states on the map. Can you think of any reasons why the EV counts in the selected state is above or below the predicted value?

11. Save the map.

Take the next step

You can share the web map and the charts you created in a web app using one of the ArcGIS Instant Apps templates called Chart Viewer. Chart Viewer allows you to quickly configure and publish a web app to display your charts alongside your map. It supports all the chart types in Map Viewer: bar, line, histogram, scatterplots, and box plots. As with Map Viewer, you can interact with the chart and map by making selections. Check the blog article at links.esri.com/Top20Charts for a step-by-step tutorial on how to use Chart Viewer.

Summary

In this chapter, you worked with US Department of Energy vehicle registration count data in CSV format. You added the data as a table to Map Viewer, exploring the attributes in the table and creating a bar chart to compare the number of electric cars in each US state. You also joined the table to a layer of US state boundaries from ArcGIS Living Atlas to visualize the data on a map and add additional attributes, including the state population. You then applied a smart-mapping style to the layer to see the spatial patterns of electric car adoption per capita. After styling the data, you created a scatterplot to analyze the correlation between the population of a state and the count of registered electric vehicles.

Workflow

1. In Map Viewer, add a layer from a file.
2. Configure settings and add the table.
3. On the Settings toolbar, configure the chart.
4. Choose a chart type and configure the chart.

CHAPTER 11
Analyzing raster layers

Objectives

- Understand what's required to perform raster analysis in ArcGIS Online.
- Learn about the differences between feature analysis tools, raster analysis tools, and raster functions.
- Calculate zonal statistics from a raster dataset using areas defined by a polygon feature layer.
- Detect and extract features from imagery using a pretrained deep learning model from ArcGIS Living Atlas.
- Customize workflows with raster functions and the Raster Function Editor (optional).

Introduction

In this chapter, you'll learn how to use ArcGIS Image to analyze a raster. Raster and imagery data contain information that can be used to identify patterns, find features, and understand change across landscapes. The increasing availability of high-resolution raster data opens up new opportunities to extract accurate information from raster datasets.

ArcGIS Image for ArcGIS Online allows you to host, manage, visualize, and analyze your imagery and raster datasets in the cloud. To run raster analysis in Map Viewer, you need to have the ArcGIS Image Online user extension (links.esri.com/ImageForOnline).

The first tutorial in this chapter uses a raster analysis tool to calculate statistics such as maximum, minimum, and mean from the United States Annual Temperature Raster within US county boundaries. This tutorial shows the statistical characteristics of normal temperature in each county in the continental United States.

The second tutorial applies a pretrained deep learning model to detect and extract objects as GIS features from high-resolution imagery. This tutorial shows how to automate feature extraction, detection, and classification from raster data such as imagery.

Requirements

- ArcGIS Image for ArcGIS Online user type extension
- ArcGIS Image Analyst privileges

Tutorial 11-1: Calculate temperature statistics in US counties

In this tutorial, you'll calculate statistics that summarize the values of the United States Annual Temperature Raster in each continental US county and generate an output table to store the multiple statistics, including maximum, minimum, and mean.

Open a map in Map Viewer and add a layer

1. Sign in to ArcGIS Online and open a new map.

   > *Review the basics for navigating the user interface in Map Viewer. The Contents toolbar is the dark vertical toolbar on the left, and the Settings toolbar is the light vertical toolbar on the right. When you click or select something on a toolbar, a pane usually opens, allowing you to adjust a related setting.*

   You'll add the hosted imagery layer, United States Annual Temperature Raster, to your map.

2. In the Layers pane, click Add.
3. In the Add layer pane, click My content and change the selection to ArcGIS Online.
4. In the search box, type **owner:Top20EssentialSkillsForArcGISOnline temperature**.

5. Locate the United States Annual Temperature Raster imagery layer, and click Add.

   The image layer is added to the map.

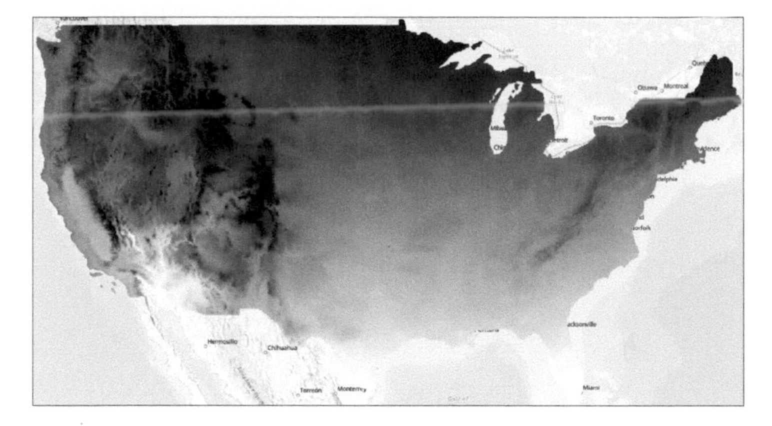

Symbolize the imagery layer

Next, you'll symbolize the temperature raster layer using the Classify option and classify the layer into regions of temperature ranges from low to high.

1. On the Settings toolbar, click Styles.
2. In the Styles pane, click Classify.

   The layer symbology updates.

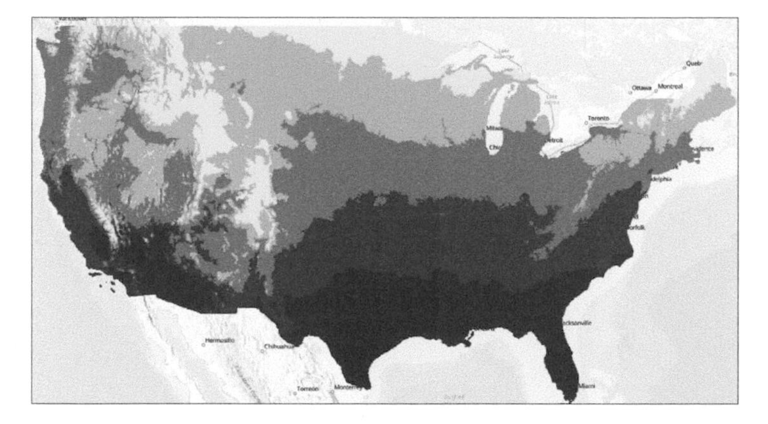

3. Under Classify, click Style options.
4. In the Style options pane, confirm that the method is Natural Breaks with five classes.
5. Click Done and click Done to close the panes.

6. On the Contents toolbar, click Legend.

   The legend shows how the layer is classified into different regions, with temperatures ranging from low (yellow) to high (red).

Perform zonal statistical analysis

You'll use the Zonal Statistics as Table analysis tool to calculate the temperature statistics such as average, maximum, and minimum in each county in the continental United States. The results of this analysis will produce a table with a summary of temperature statistics by county boundary.

1. On the Settings toolbar, click Analysis.

   > **Warning:** *If you don't see the Analysis button, contact your ArcGIS administrator. You may not have the privileges required to perform analysis.*

2. In the Analysis pane, click Tools.

   Tools in Map Viewer include feature analysis tools and raster analysis tools. You can differentiate a feature tool from a raster tool based on the different icons. The hammer icon on top indicates a feature tool, and the hammer-and-squares icon below indicates a raster tool.

   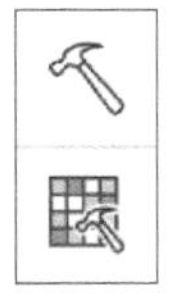

   Next, you'll filter the tools based on the analysis type to show only the raster analysis tools.

3. On the Tools pane, click Filter and click Raster analysis.

> **Warning:** *If you don't see the Raster Analysis option, contact your ArcGIS administrator. You may not have the ArcGIS Image Online user type extension or the privileges required to perform raster analysis.*

Raster analysis in ArcGIS Online includes raster analysis tools and raster functions. Raster analysis tools apply operations to raster datasets and generate new output as hosted layers or tables. Raster functions are dynamic operations that apply processing to the pixels of raster datasets without creating new output data. You can combine raster functions into a processing chain as a raster function template. You can also tailor raster function templates for many applications using a variety of input data types and raster functions to facilitate specific workflows.

4. Under Summarize data, click Zonal Statistics as Table.

   Next, you'll choose a layer to define the zones that will be summarized.

5. Under Input zone raster or features, click Layer.

   Once you identify this layer, the tool will use it to calculate temperature statistics within each county.

6. In the Select layer pane, at the bottom, click Browse layers.

   In this case, you'll choose a layer of US county boundaries from ArcGIS Living Atlas.

7. Click My content and change the selection to Living Atlas analysis layers.

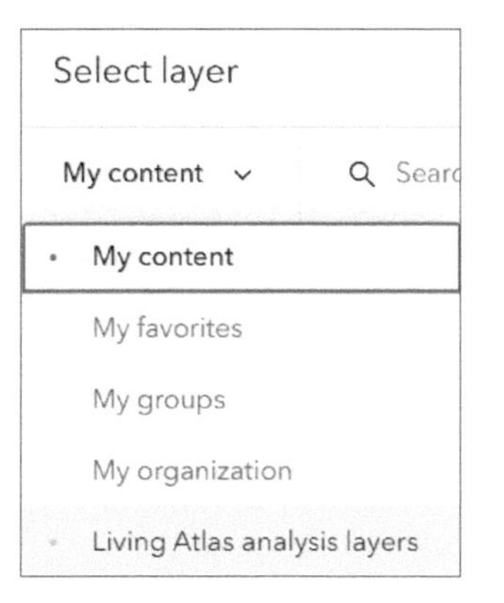

8. In the search box, type **USA Counties**.

9. Locate the USA Counties Generalized Boundaries layer from the search results.

10. Click Select layer and click the layer name. At the bottom of the window, click Confirm.

> **Hint:** By clicking *Confirm* instead of *Confirm and add to map*, you can run the analysis from a public layer without the extra step of adding it to the map.

11. In the Zonal Statistics as Table pane, under Zone field, click OBJECTID and click Name.

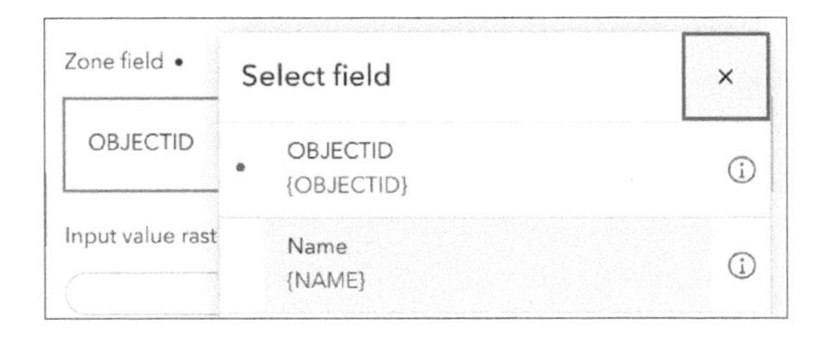

12. Under Input value raster, click Layer. In the Select layer pane, click United States Annual Temperature Raster.
13. In the Statistical analysis settings section, under Statistic type, click All and choose the combined option for Minimum, Maximum and Mean.
14. Under Output table name, type **TemperatureStatisticsByCounty <your name>**.
15. Under Save in folder, click the arrow and choose Top 20 Tutorial Content.

> *If you need to create the folder, click the down arrow and choose Create new folder. Type* **Top 20 Tutorial Content**.

16. Under Environment Settings, click the arrow.

In the Environment settings, you'll indicate that you want the extent of a layer to limit the processing extent of the analysis.

> **Hint:** Limiting the processing extent of your analysis can help you reduce the number of credits used.

17. Under Processing extent, click the arrow and choose Layer.

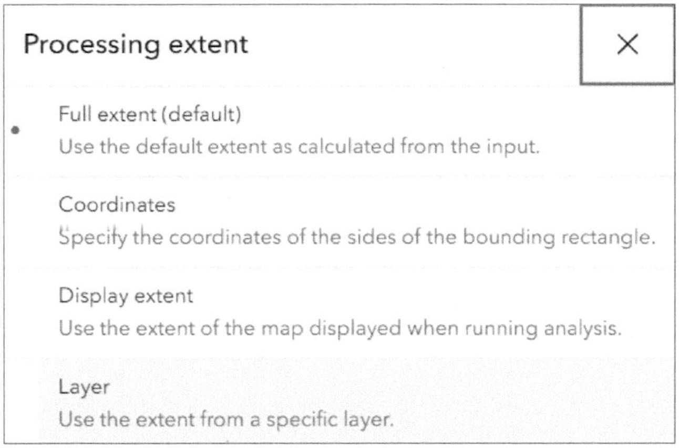

Processing extent ×

Full extent (default)
Use the default extent as calculated from the input.

Coordinates
Specify the coordinates of the sides of the bounding rectangle.

Display extent
Use the extent of the map displayed when running analysis.

Layer
Use the extent from a specific layer.

Next you will specify which layer's extent will be used.

18. Under the Layer extent selection, click the Layer button. In the Select layer pane, choose United States Annual Temperature Raster layer.

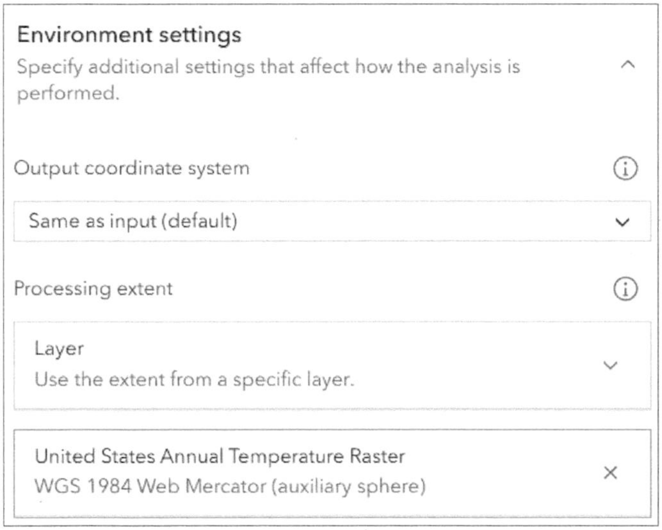

Environment settings
Specify additional settings that affect how the analysis is performed.

Output coordinate system

Same as input (default)

Processing extent

Layer
Use the extent from a specific layer.

United States Annual Temperature Raster
WGS 1984 Web Mercator (auxiliary sphere)

Environment settings can be applied to a web map or an individual tool. To learn more about environment settings, go to **doc.arcgis.com**, *type* **ArcGIS Online** *in the search box to filter your search, click the ArcGIS Online chiclet, click Analyze on the black ribbon, expand Overview in the topic tree, and click Set Analysis Settings.*

19. At the bottom of the tool, click Estimate Credits.

Running the tool will use approximately one credit.

20. Click Run.

21. At the top of the tool pane, click the History tab.

As the tool runs, you can view the progress.

> *Each time you run a tool in a web map, information about the tool and its parameters is saved in History. You can view detailed information about each job and open and run any of the tools in History, modifying the previous parameters. To learn more about analysis history, go to* **doc.arcgis.com**, *type* **ArcGIS Online** *in the search box to filter your search, click the ArcGIS Online chiclet, click Analyze on the black ribbon, expand Overview in the topic tree, and click Analysis History.*

When the tool has completed running, the table is added to your map in the Tables pane.

22. On the Contents toolbar, click Tables.

23. In the Tables pane, click TemperatureStatisticsByCounty to view the table.

TemperatureStatisticsByCounty
1,843 records

NAME	ZONE_CODE	COUNT_	ZONE_AREA	MIN_	MAX_
Autauga County	1	3,422.00	2,190,080,000.00	17.03	18.62
Baldwin County	2	10,683.00	6,837,120,000.00	17.36	20.16
Barbour County	3	7,337.00	4,695,680,000.00	8.10	18.74

The Zonal Statistics as Table tool calculated each US county's minimum, maximum, and mean temperature.

24. Sort the columns in ascending or descending order and answer the following questions:

- Which county has the highest temperature recorded?
- Which county has the lowest temperature recorded?
- Which county has the highest average temperature?

Save your map

Next, you'll save the web map and assign it a title, tags, and a summary so it's easy to find later.

1. In the Contents toolbar, click Save and open > Save as.
2. In the Save map window, for Title, type **Calculate Zonal Statistics <your initials>**.
3. For Folder, choose Top 20 Tutorial Content.
4. For Tags, type **Top 20, ArcGIS Online, imagery, temperature**.
5. For Summary, type **USA Temperature Statistics by County**.
6. Click Save.

Tutorial 11-2: Detect and extract features from imagery using deep learning

Building and training your own deep learning models or fine-tuning existing trained models is an advanced task. You would need large volumes of training data and imagery, computer resources, and the data science expertise to train such models. Esri provides a growing library of pretrained deep learning models on ArcGIS Living Atlas. By using these models, you can get started right away with using artificial intelligence to extract information and gain insights from your imagery.

In this tutorial, you'll use deep learning analysis in ArcGIS Image Online to extract cars from a satellite imagery layer and turn them into feature data that can be used in a GIS workflow.

> *Deep learning is a subset of machine learning, which is a form of artificial intelligence. ArcGIS has tools that use the latest innovations in deep learning, including image classification—assigning a label to a digital image. For example, an aerial drone image might be labeled "crowd," or a photo of a cat might be labeled "cat." This classification can then be used in GIS to categorize other features in an image.*

Add an imagery layer to a new web map

> *If you didn't finish the previous tutorial, sign in to ArcGIS Online, open a new blank map, and start with the second step.*

1. In Map Viewer, on the Contents toolbar, click Save and open > New map.

   Now you'll add a hosted imagery layer to your new map.

2. In the Layers pane, click Add.
3. In the Add layer pane, click My content and change the selection to ArcGIS Online.
4. In the search box, type **owner:Top20EssentialSkillsForArcGISOnline Ventura**.
5. In the search results, find the Ventura Imagery and click Add.

   The image layer is added to the map.

   Zoom and pan around the map to inspect the car locations.

6. At the top of the Layers pane, click the back arrow.
7. In the Layers pane, at the end of the Ventura Imagery layer, click Options and click Zoom to.

   The web map zooms to the extent of the Ventura Imagery layer.

Save the web map

Before continuing, you'll save your web map and assign it a title, tags, and a summary to make it easy to find and identify later.

1. On the Contents toolbar, click Save and open > Save as.
2. In the Save map window, for Title, type **DeepLearningExtractCars <your initials>**.
3. For Folder, choose Top 20 Tutorial Content.
4. For Tags, type **Top 20, ArcGIS Online, imagery, Ventura**.
5. For Summary, type **Detect and extract features from imagery using deep learning.**
6. Click Save.

Extract features from imagery

There are thousands of cars in this image, but you don't have much data about them. You don't know how many there are or the location of each one. To find out, you could manually create a point for each car and store the features in a feature layer, but this would be tedious and time-consuming. So you'll use artificial intelligence and a pretrained deep learning model from ArcGIS Living Atlas to extract information from the image and provide it in a new layer.

1. On the Settings toolbar, click Analysis.
2. In the Analysis pane, click Tools.
3. In the search box, type **deep learning** and click Detect Objects Using Deep Learning.

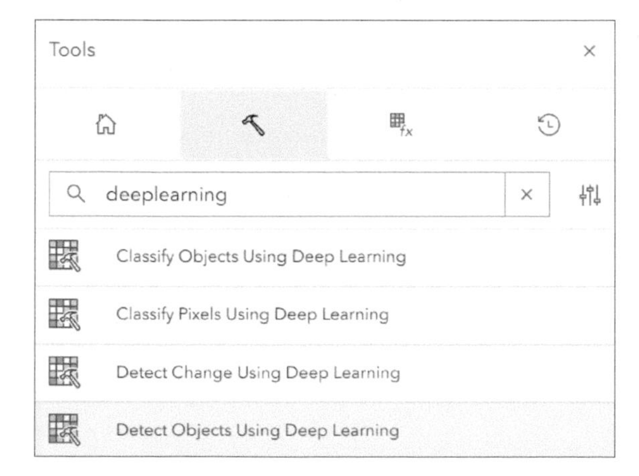

4. In the Detect Objects Using Deep Learning tool pane, under Input imagery layer or feature layer, click Layer.
5. In the Select layer pane, click Ventura Imagery.
6. Under Model for object detection, click Select model.

   The Select item window appears.
7. Click My content and click Living Atlas. In the search box, type **cars**.
8. Click Car Detection – USA.

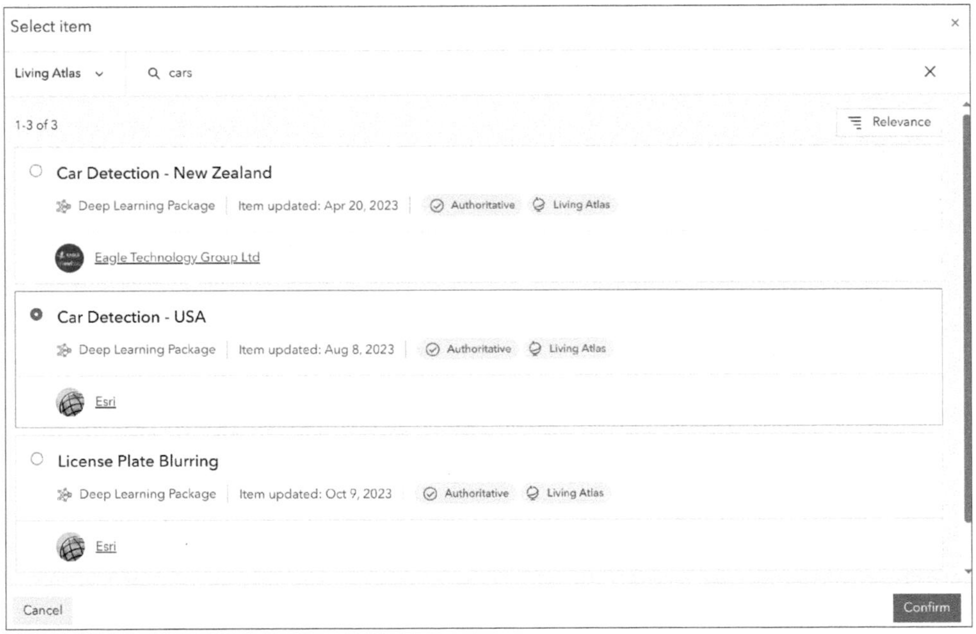

9. Click Confirm.

   Use the default model arguments.
10. In the tool pane, under Output name, type **Cars_Ventura <your initials>**.
11. Under Save in folder, choose Top 20 Tutorial Content.

On your own

In Environment Settings, set the extent to the Ventura Imagery layer, just as you did in the previous tutorial.

12. Click Estimate Credits.

    The deep learning operation will use 4.8 credits.

13. Click Run.

14. At the top, click the History tab.

    The tool may take 15 to 20 minutes to run.

    When the tool completes, the layer is added to the Layers pane. On the map, a feature layer shows the extracted cars on top of the imagery layer.

15. On the map, zoom in to inspect the results of the analysis.

    The deep learning model did a good job of detecting cars in the imagery and created polygons from them.

16. In the Layers pane, click the Cars_Ventura layer. Click the options button for the layer, and click Show table.

    The attribute table for the Cars_Ventura layer has a column called Confidence. This field indicates how confident the deep learning model is that each group of pixels in the image is a car.

Confidence
98.68
97.49
98.79

17. In the Contents pane, click Save and open > Save.

Take the next step

In tutorial 11-1, you created an output table. Use the Join Features analysis tool to join the output table to the USA counties boundaries layer in ArcGIS Living Atlas. Create a web map to visualize temperature patterns for US counties.

Summary

In this chapter, you used the Zonal Analysis as Table tool to calculate summary statistics of temperature in each US county. In tutorial 11-2, you ran a pretrained deep learning model to detect and extract cars from high-resolution imagery.

To learn more about raster functions and raster function templates, search Esri online documentation. Go to doc.arcgis.com, type **ArcGIS Online** in the search box to filter your search, click the ArcGIS Online chiclet, click Analyze on the black ribbon, expand Map Viewer Classic in the topic tree, expand Raster Analysis, and expand Raster Functions to view the "Raster Functions" and "Raster Function Editor" topics.

Workflow

1. Add an imagery layer.
2. Symbolize the layer.
3. Choose and configure a raster analysis tool.

CHAPTER 12
Analyzing feature layers

Introduction

Together with mapping, spatial analysis is a fundamental component of ArcGIS Online. Spatial analysis goes beyond map visualization to answer questions about spatial relationships and patterns. Whether it's overlaying layers to identify areas of interest, conducting proximity analysis to understand spatial relationships, or detecting and quantifying patterns to find hot spots and outliers, the analysis tools in Map Viewer reveal insights and use spatial information for problem-solving.

Tutorial 12-1: Analyze locations and populations at risk

In 2018, more than 20 fissures opened in the Puna district, Hawaii, spewing magma and poisonous gas. Hundreds of residents were evacuated, and more were forced to leave their homes.

In this tutorial, you'll use feature analysis to predict the Kilauea lava flow impacts on shelter access. To estimate the affected areas, you'll create a buffer around the potential lava flows and refine it to show only areas on land.

Then you'll determine how many people live in the affected area. In the second tutorial, you'll analyze volcano shelters and create drive-time areas to find out how the lava flow path could cut off accessibility to shelters.

Open a web map and add data

1. In a browser, sign in to ArcGIS Online, and open a new, blank map.

   First, you'll add a layer of lava vents to identify where lava can come from.

2. On the Contents toolbar, in the Layers pane, click Add.

   > *Review the basics for navigating the user interface in Map Viewer. The Contents toolbar is the dark vertical toolbar on the left, and the Settings toolbar is the light vertical toolbar on the right. When you click or select something on a toolbar, a pane usually opens, allowing you to adjust a related setting.*

3. In the Add layer pane, click My content and click ArcGIS Online.

4. In the search box, type **owner:Top20EssentialSkillsForArcGISOnline Fissures2018**.

5. In the search results, under Fissures2018, click Add.

   The layer is added to the map, showing volcanic activity along the coast of Hawaii's Big Island.

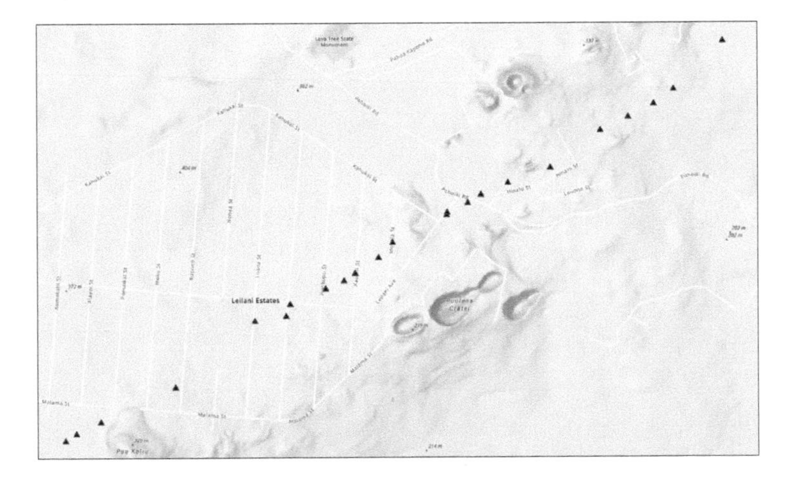

6. On the Add layer pane, at the top, click the back arrow to return to the Layers pane.

Save the web map

Next, you'll save the web map and provide a title, tags, and a summary to make it easy to find later.

1. On the Contents toolbar, click Save and open > Save as.

2. Save your map with the following settings:

   - Title: **Lava impact and shelter access <Your Initials>**
   - Folder: Top 20 Tutorial Content

   > *If you need to create the folder, click the down arrow and choose Create new folder. Type* **Top 20 Tutorial Content**.

   - Tags: **Top 20, lava, analysis**
   - Summary: **A map analyzing lava flows and shelter access in Hawaii.**

3. Click Save.

Predict the potential flow path of the lava

Lava, like water, flows downhill along the path of least resistance. Because of this, you can use the Trace Downstream analysis tool to predict its behavior.

1. On the Settings toolbar, click Analysis.

2. In the Analysis pane, click Tools.

   Map Viewer includes feature analysis tools. If you have the ArcGIS Image for ArcGIS Online user type extension, you will also have raster analysis tools. To find a tool, you can filter based on analysis type or search using a keyword. You can also browse and select a tool from preset categories.

3. In the Tools pane, in the search box, type **Trace**. From the search results, click the Trace Downstream tool.

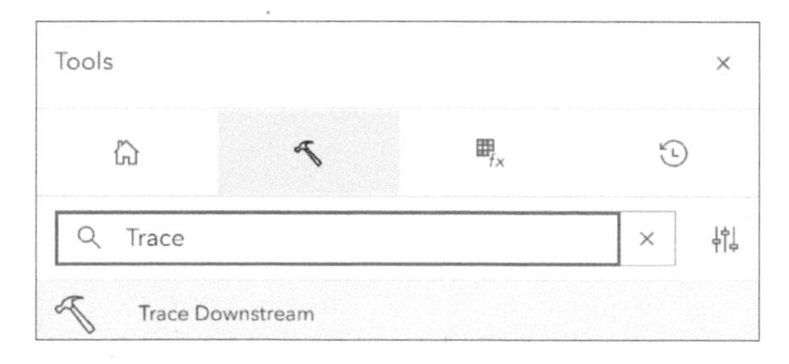

4. In the Trace Downstream tool pane, under Input point features, click Layer and choose the Fissures2018 layer.

   The input layer is set.

   Input point features •

   Fissures2018
   Count of features: 23

5. Accept the remaining default values for the Downstream path settings.

   You can change the maximum length of the paths and units and clip the output.

6. Under Output line name, type **Potential Lava Flows <Your Initials>**.

7. Under Save in folder, choose Top 20 Tutorial Content.

> **Hint:** Most analysis tools consume credits when they run, but the Trace Downstream tool doesn't, so you don't need to click Estimate credits.

8. Click Run.

   To check the progress, at the top of the pane, click History.

   A message indicates that the process has completed successfully, and the layer appears on the map.

9. Zoom out slightly to review the results.

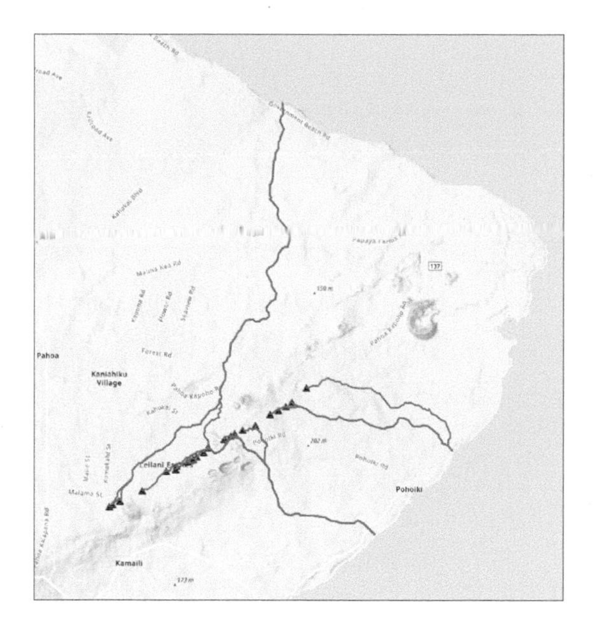

The Trace Downstream tool drew possible lava flow routes to the north, east, and southeast of the vents. Based on these routes, the lava has the potential to cut off a portion of Puna district and could be devastating to the community.

10. Save the map.

Estimate potentially affected areas

Next, you'll create a buffer around the Potential Lava Flows layer to estimate the affected areas.

1. At the top of the Trace Downstream tool pane, click the back arrow to return to the Tools pane.

2. In the Tools pane, in the search box, type **Buffer**. From the search results, click Create Buffers.

3. In the Create Buffers tool pane, under Input layer, click Layer and click Potential Lava Flows.

> *The Potential Lava Flows layer is the result of running the Trace Downstream tool, which you created in the previous section.*

4. For Distance type, confirm that Value is selected.

5. Under Distance values, type **1** and click Add.

   A 1 is added.

6. Under Units, click the list and click Miles.

7. Under Overlap policy, click the list and click Dissolve.

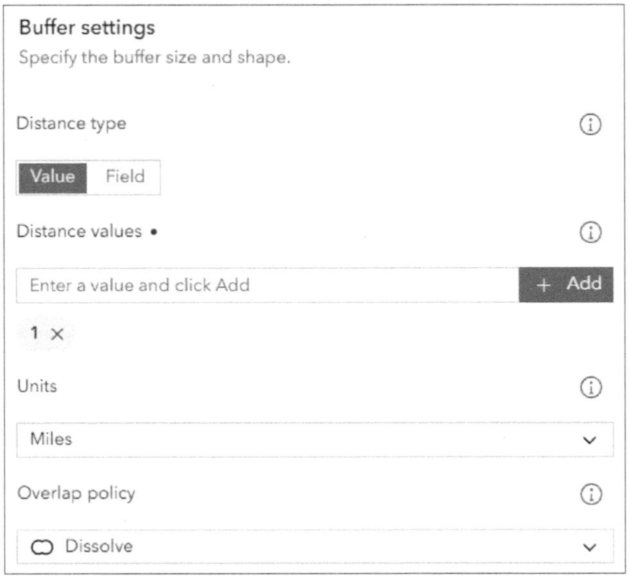

   The tool is configured to create a one-mile buffer around the potential lava flow area.

8. Under Output name, type **Affected Areas <Your Initials>**.

9. Under Save in folder, click the list and click Top 20 Tutorial Content.

10. Click Estimate Credits.

11. Click Run.

    The tool creates a buffer.

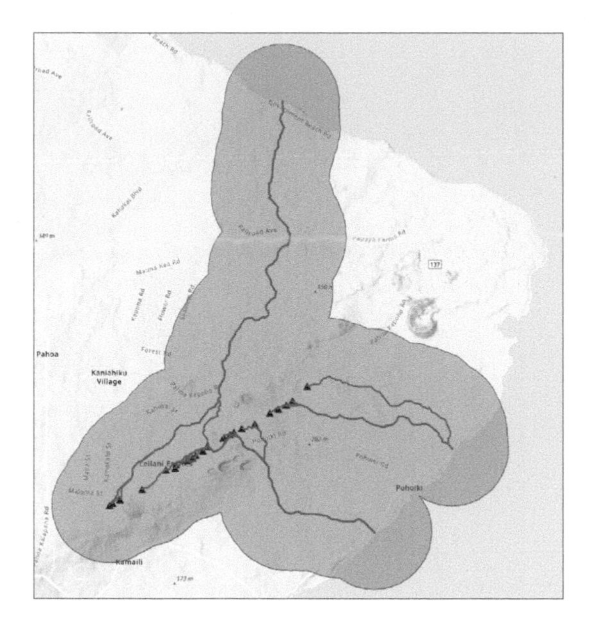

The buffer includes ocean areas that are irrelevant to the analysis.

Refine your estimate of affected areas

Next, you'll use the Overlay Layers tool to clip the layer so that only the land area is analyzed.

1. Return to the Tools pane.

> **Hint:** At the top, click the back arrow.

2. In the Tools pane, clear the previous input in the search box.
3. In the tools list, expand the Manage data category and click Overlay Layers.
4. In the Overlay Layers pane, under Input features, click Layer and choose Affected Areas.

   The Affected Areas layer is the result of running the Create Buffers tool.

5. Under Overlay features, click Layer. At the bottom of the Select layer pane, click Browse layers.
6. In the Select layer window, click My content and click Living Atlas analysis layers.
7. In the search box, type **county**.

8. Locate the United States County Boundaries layer in the search results, click Select layer, and click USA_County.

Living Atlas analysis layers
county
1-20 of 34
Relevance
Date modified
Date created
Item type
Tags
Sharing
Status
USA Counties Generalized Boundaries
Feature Layer | Item updated: Mar 2, 2023
Esri
Select layer
United States County Boundaries
Feature Layer | Item updated: May 23, 2023
Esri
Selected layer: USA_County
USA_County

9. Click Confirm and click Add layer to map.

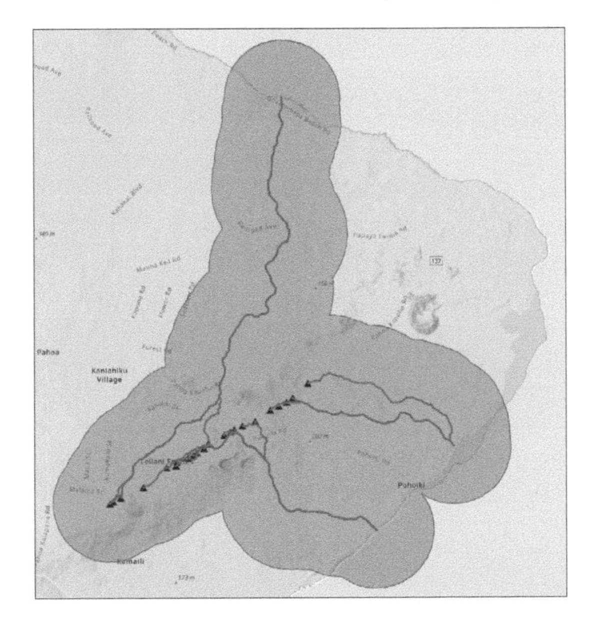

10. In the Overlay Layers tool pane, under Overlay type, confirm that Intersect is selected.

11. For Output name, type **Affected Areas on Land <Your Initials>**.

12. Under Save in folder, click the list and choose Top 20 Tutorial Content.

13. Click Environment settings. Under Processing extent, click the list and choose Display extent.

> **Hint:** Environment settings can be applied to the web map or a tool, as shown here. Check ArcGIS Online documentation to learn more about environment settings for Map Viewer analysis.

14. Click Estimate Credits.

15. Click Run.

16. When the layer is added to the Layers pane, point to the Affected Areas layer to see the visibility icon (eye) and turn off layer visibility.

On your own

Turn off the USA County layer visibility.

The result shows only the land area that might be affected by lava flows.

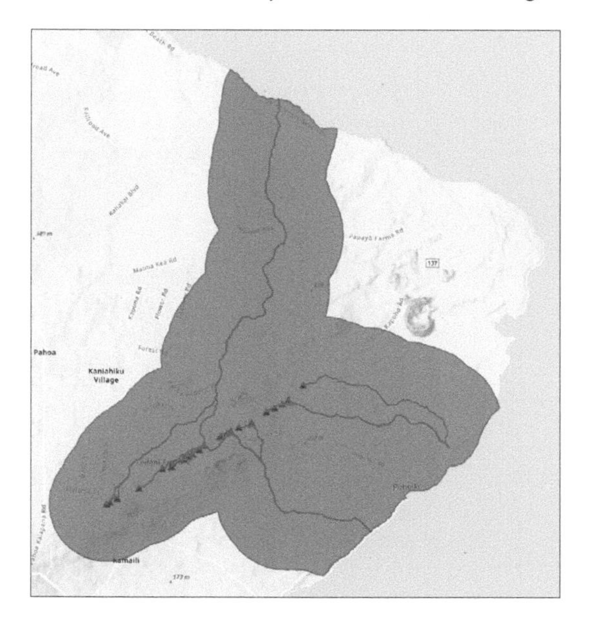

17. Save the map.

Estimate the affected population

Next, you'll use the Enrich Layer tool to find out how many people live in this area and might be affected by a lava flow.

The Enrich Layer tool improves your spatial feature data with demographic, socioeconomic, and landscape information. The service provides more than 15,000 data fields for locations in more than 150 countries and regions.

1. Return to the Tools pane.
2. In the Tools pane, expand Enrich data and click Enrich Layer.
3. In the Enrich Layer pane, for Input features, choose Affected Areas on Land.
4. For Enrichment variables, click Variable.
5. In the Data Browser window, click Population.
6. In the search box, type **2018** and press Enter.
7. Select 2018 Total Population (Esri 2023).

   The variable is added to the Selected Variables count in the upper right.

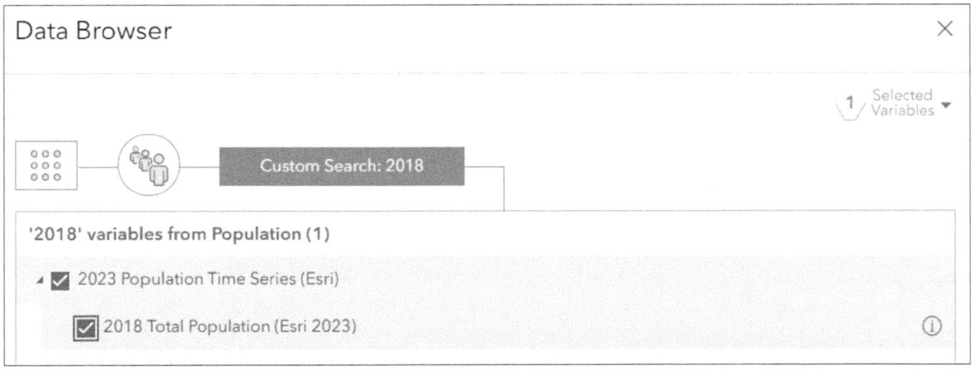

8. In the lower right, click Select.
9. In the Enrich Layer tool, for Output name, type **Affected Population <Your Initials>**.
10. For Save in Folder, select Top 20 Tutorial Content.
11. Click Estimate Credits.

12. Click Run.

    When processing is complete, the layer is added to the map.

13. Return to the Tools pane.

14. In the Layers pane, point to the Affected Population layer and click Options > Show table.

    The Affected Population attribute table appears.

15. In the table, locate and expand the 2018 Population column, which was added by the Enrich Layer tool.

2018 Population (Esri 2023)
2,966.00

    This number indicates the number of people who live in the affected area.

16. Close the table and save your map.

Tutorial 12-2: Analyze volcano shelter access in Hawaii

Now, you'll analyze volcano shelter accessibility in Hawaii. You'll identify which areas currently have access to volcano shelters and identify areas on the island that could be cut off by lava flows, thus preventing public access to shelters.

Identify shelter locations

First, you'll add a layer showing where the emergency shelters are.

1. In the Layers pane, click Add.

2. In the Add layer pane, click ArcGIS Online, and in the search box, type **owner:Top20EssentialSkillsForArcGISOnline island shelters**.

3. On the resulting Island Emergency Shelters layer, click Add.

    The layer is added to the map showing the locations of emergency shelters on the island.

4. On the Add layer pane, click the back arrow.
5. In the Layers pane, point to the Island Emergency Shelters layer, and click Options > Zoom to.

   The map zooms to the extent of the Emergency Shelters layer.
6. In the Layers pane, turn off layer visibility for all layers except the Affected Population layer and the Island Emergency Shelters layer.

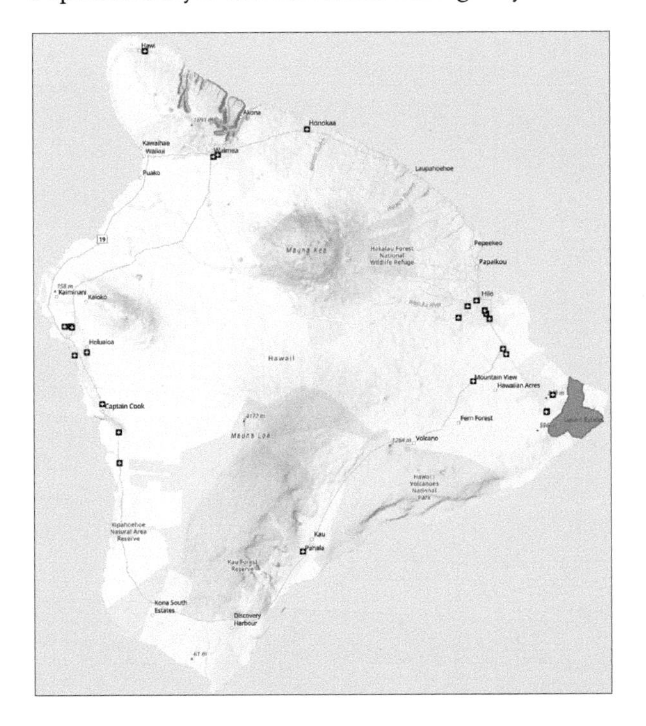

Define the analysis processing extent

Next, you'll focus on the area potentially affected by lava flows, which you identified previously.

1. Zoom to the extent of the Affected Population layer.

> **Hint:** In the Layers pane, point to the Affected Population layer, and click Options > Zoom to.

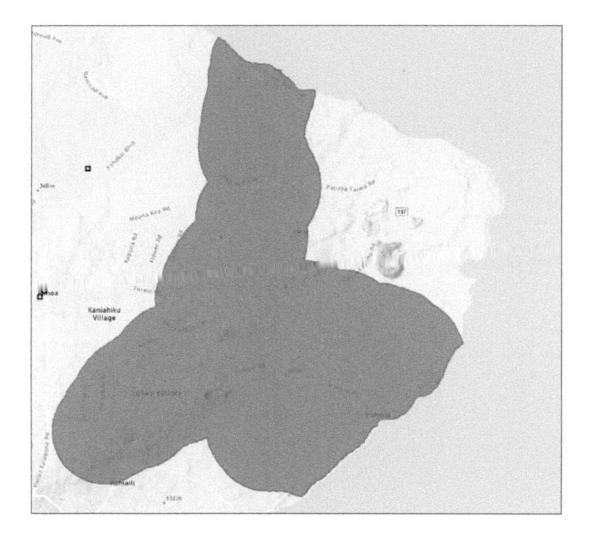

Now you'll set the current map extent as the processing extent for your analysis.

2. At the top of the Tools pane, click Home.

3. At the bottom of the Analysis pane, click Analysis Settings.

4. Under General Environments, under Processing extent, choose Display extent.

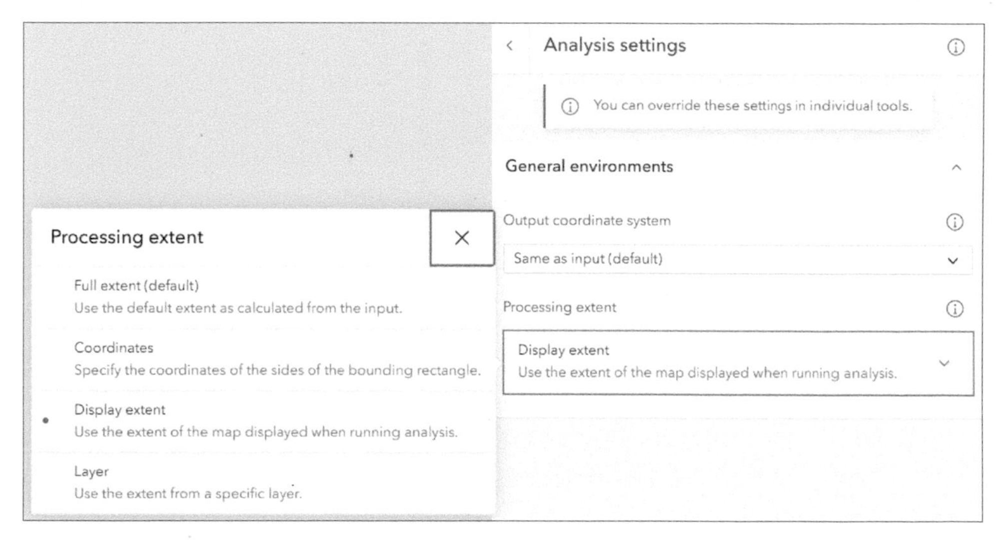

5. In the Analysis settings pane, click Save.

This setting applies to the web map, which means that any tool you open in this web map will use this processing extent by default. Before running a tool, you still have the option to overwrite the map-level settings under the tool's environment settings option.

6. Save your map.

Analyze access to shelters

Next, you'll analyze accessibility by creating drive-time areas around the emergency shelters.

1. Turn off the Affected population layer visibility.
2. In the Analysis pane, click Tools.
3. In the Tools pane, search for and open the Generate Travel Areas tool.

   The Generate Travel Areas tool calculates the area that can be reached within a specific travel time or distance along a street network. In this case, you'll calculate a 15-minute drive to shelters. Shelters within this distance can be considered easily accessible.

4. For Input layer, choose Island Emergency Shelters.
5. For Travel mode, confirm that Driving Time is selected.
6. For Cutoffs, type **15** and click Add.
7. Confirm that Cutoff units are set to Minutes.
8. For Travel direction, choose Toward input locations.

   > *Choose this selection because people will be traveling toward the shelters.*

9. Leave the Arrival time set to Time unspecified.
10. For Overlap policy, choose Dissolve.

    The default Overlap option creates a unique drive-time area for each shelter. Dissolving the areas creates one output feature instead of many. The same total area will be covered.

11. For Output name, type **Shelter in 15 Minutes <your initials>**.
12. For Save in folder, choose Top 20 Tutorial Content.

13. Click Estimate credits.

    This analysis will use 1.5 credits.

> **Hint:** If your estimated credits show a higher number, double-check the Environment settings and confirm that the processing extent is set to Display Extent.

14. Click Run.

    The layer is added to the map.

The layer shows the areas within the analysis extent where people can drive to an emergency shelter in 15 minutes or less.

15. Save the map.

Identify areas that might be cut off from shelter access

Lava flows may prevent access to shelters. Next, you'll use the Generate Travel Areas tool again to create 15-minute drive-time areas to emergency shelters. This time, you'll analyze access after adding the potential lava flows as a travel barrier.

> **Hint:** Barriers can be added to network analysis tools such as Generate Travel Areas and Find Closest to consider temporary changes on a street network. For example, if there is a temporary road closure because of a fallen tree, accident, or flood, you can apply street barriers when running routing-based analysis tools.

To use the analysis history feature, you will rerun the last workflow with a change to the parameter settings.

1. At the top of the Analysis pane, click History.

   In Map Viewer, each time you run a tool in a web map, information about the tool and its parameters is saved in history. All tool runs are added to the analysis history, including successfully run, failed, or canceled jobs. You can reopen and run any of the tools in history with the same or modified parameters.

2. In the History pane, next to the Generate Travel Areas, click the more button and click Open tool.

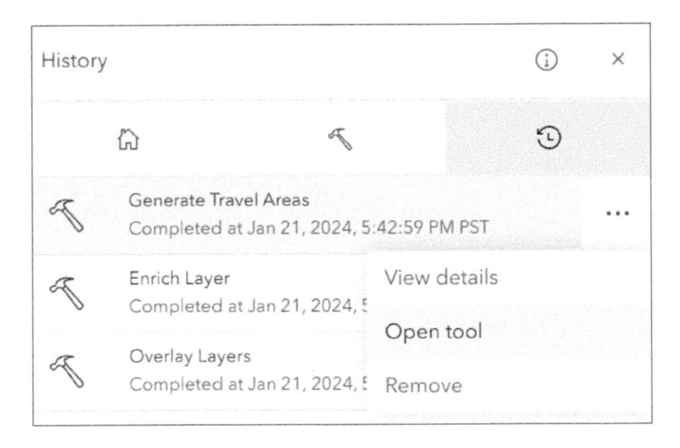

   The tool opens with the previous settings applied.

3. Expand the Optional barrier layers heading.
4. For Line barrier layer, choose Potential Lava Flows.
5. For Output name, type **Shelter in 15 minutes with barrier <Your Name>**.
6. Click Estimate credits.
7. Click Run.

   The layer is added to the map.

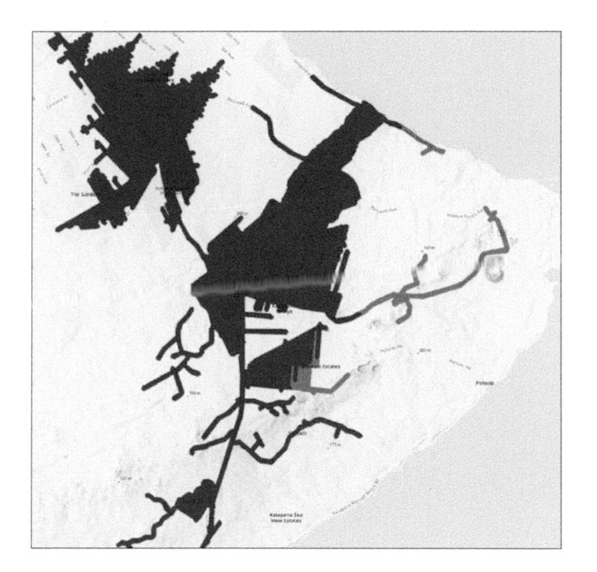

The new layer, in dark purple, shows the areas within a 15-minute drive of a shelter using the potential lava flows as a barrier. The areas visible in the previous layer (light purple) can no longer access the shelters in 15 minutes.

8. Save your map.
9. On the Contents pane, click Share map and share your map with Everyone. Update sharing to include all layers in the map.

Take the next step

Style the map to illustrate your findings. Turn on layer visibility for the Fissures2018, Potential Lava Flows, Affected Population, and Island Emergency Shelters layers. Order the layers in the Layer pane as appropriate. Style the two travel-time area layers differently to show the impact on shelter access. Style the other layers as you think best to communicate your message.

Summary

In this chapter, you performed feature analysis to quantify patterns and relationships. First, you used the Trace Downstream, Create Buffers, Overlay Layers, and the Enrich Layer tools to predict lava flow and estimate how many people may be affected. Next, you analyzed volcano shelter accessibility by applying network analysis to generate drive-time areas around the emergency shelters.

CHAPTER 13

Adding a media layer

Objectives

- Add a media layer to a basemap.
- Position the media layer.
- Enhance the appearance of the media layer.

Introduction

Media layers are images overlayed on a map. Often made by uploading scanned paper maps or aerial images, media layers are positioned in the correct location to show the image in the context of a map. By adding a media layer on a basemap, you can compare the image with a current map of the area.

Tutorial 13-1: Add a media layer, position it, and adjust its appearance

In this tutorial, you'll download an image of a survey, completed in 1862, of Seattle, Washington. You'll add an image of the survey to a map as a media layer. Then you'll rotate, resize, and position the image in the correct geographic location. Finally, you'll improve the appearance and review the historical survey in the context of the current basemap.

Download an image

Download an image file of the historical survey to your computer.

1. Sign in to ArcGIS Online.
2. On the top navigation bar, click the magnifying glass to search for items in ArcGIS Online.
3. Type **owner:Top20EssentialSkillsForArcGISOnline survey Seattle** and press Enter.
4. Under Filters, turn off the filter for your organization and expand your search to all of ArcGIS Online.

   The Survey Plat Seattle item appears in the search results. You'll use this item to create a media layer.

5. On the Survey Plat Seattle item, right-click Download.
6. Click Save link as. Browse to a convenient location on your computer to store the image, and click Save.

Open a new, blank map in Map Viewer and add a media layer

You'll add a media layer in a new map.

1. On the top navigation bar, click Map to open a new, blank map.

   > *Review the basics for navigating the user interface in Map Viewer. The Contents toolbar is the dark vertical toolbar on the left, and the Settings toolbar is the light vertical toolbar on the right. When you click or select something on a toolbar, a pane usually opens, allowing you to adjust a related setting.*

2. In the Layers pane, click the down arrow next to Add and choose Add media layer.
3. In the Add media pane, drag the image you downloaded, or click Your device, browse to the file, and add it.

   The Add media layer pane has a semitransparent survey image overlaid on the map.

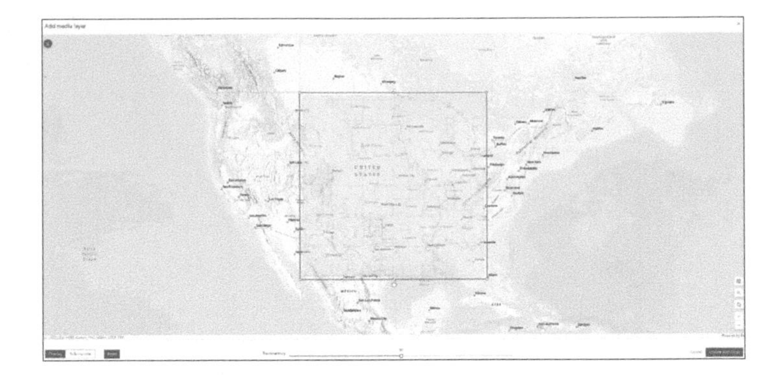

Adjust the location and scale of the media layer

This image is a simple media file that doesn't have any spatial metadata you can use to place it in a location. You'll reposition the image and manually adjust the size to place it in the correct spot.

1. In the lower right, click the magnifying glass to start a search.

   The search bar opens in the upper right.

2. In the search bar, type **Seattle, WA** and press Enter.

   The map zooms to Seattle.

3. In the search result pop-up, click Reposition media here.

   The survey is repositioned over the Seattle area.

4. Close the pop-up.
5. At the bottom of the window, drag the transparency slider from 0 to 100 and back a few times to get a sense of the area in the survey and the corresponding location in the basemap. Consider what adjustments you'll need to make to line up the survey with the basemap.
6. Set the transparency slider to 25 so you can see the survey and the basemap.

   The adjustments you'll need to make may vary, but you'll probably need to make the survey smaller and move it to the south. You'll use the orange handles to make some adjustments.
7. While holding down the Shift key, click one of the corner handles and drag the corner to make the survey image smaller while maintaining the aspect ratio. Release the corner.
8. Find some reference points, such as the peninsula on the eastern coast, that you can use as a guide for finding the correct size and location for the survey image. Drag to reposition the survey image over the basemap.

   The landscape has some significant changes, especially the western side, where tidelands have been filled in and the coastline has expanded.

   > **Hint:** At the upper left, click the lightbulb to see shortcuts you can use to position the element.
9. Zoom in and continue to resize the survey image, refining the placement until you've found a close fit.

   > **Hint:** If you want to restart the repositioning process, click Reset at the bottom of the screen.

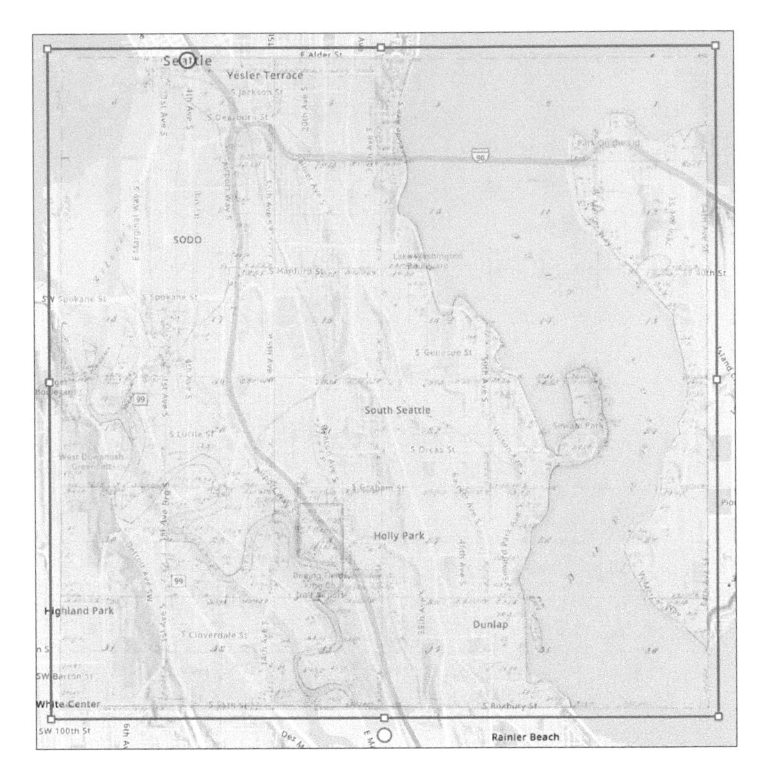

10. When you've finished, click Update and close.

Add a layer to a basemap

The survey is based on the Public Land Survey System, which gives the survey its characteristic squares and grid pattern. Next, you'll add a layer from ArcGIS Living Atlas that shows those survey lines. Survey lines make it easy to fine-tune your overlay by lining up the lines with your survey.

1. On the Contents toolbar, click Basemap and click Light Gray Canvas.
2. At the top of the Basemap pane, click Current basemap Light Gray Canvas.

3. Under Reference, point to the Light Gray Reference layer and click the visibility button to turn off the layer labels.

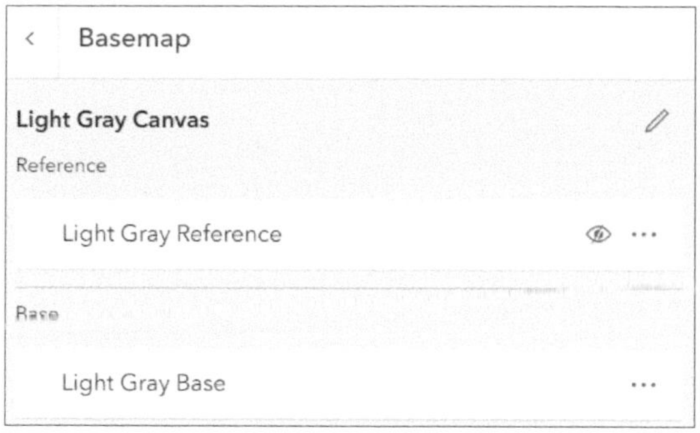

< Basemap

Light Gray Canvas

Reference

Light Gray Reference

Base

Light Gray Base

With the labels off, you can easily focus on the landform. You'll add to your basemap a layer that has Public Land Survey System lines.

4. In the Basemap pane, click Add.
5. In the Add layer pane, click My content, and click ArcGIS Online.
6. In the search box, type **owner:Top20EssentialSkillsForArcGISOnline PLSS**.
7. In the search results, under PLSS Sections, click Add.

The layer is now part of your basemap, and the survey lines can be used to position your media.

Position your media in side-by-side mode

To place the survey in the map, you used *overlay mode*, but now you'll use *side-by-side mode,* which is especially useful for fine-tuning positions. Side-by-side mode lets you place pins where you want corresponding locations in your images to be positioned. To do this, you'll need to identify the same point in the media and the map, which can be a building, intersection, or other landmark. For now, you'll use survey lines.

1. On the Contents toolbar, click Layers and choose Survey.
2. In the Properties pane, under Information, click Edit placed media.
3. At the bottom of the screen, click Side-by-side.

Now you can reposition your layer using pins in side-by-side mode.

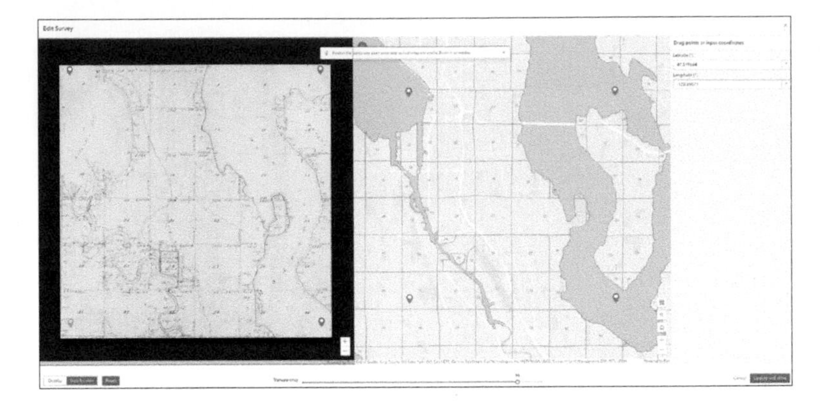

4. On the map, zoom to the pin in the upper left.
5. Click the pin and drag it to the intersection of sections 05, 04, 08, and 09.
6. Zoom in to place the point precisely.

7. In the survey, click the upper-left pin and place the point at the same intersection of sections 05, 04, 08, and 09. Zoom in as needed.

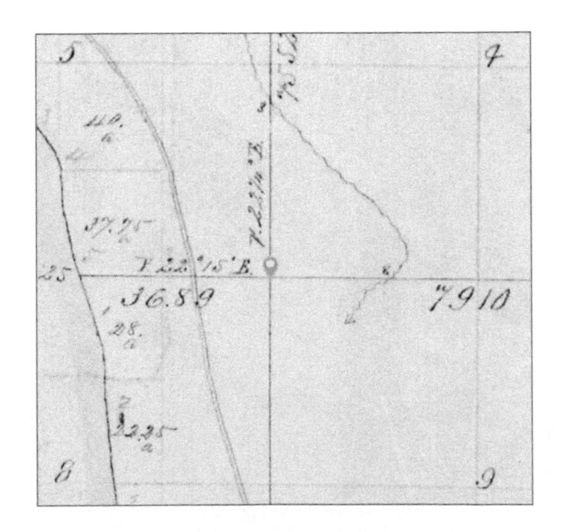

8. Repeat the process for the other three pins. Choose intersections of section lines that are easy to find in the map and the survey. You might try the following points:

   - Upper right: intersection of sections 02, 01, 11, 12
   - Lower left: intersection of sections 30, 29, 31, 32
   - Lower right: intersection of sections 27, 26, 34, 35

9. When you've finished, click Update and close.

Adjust the appearance

The Properties pane has several settings to make your media layer stand out.

1. On the Contents toolbar, click Basemap and choose Human Geography Map.

2. At the top of the list of basemaps, click Current basemap, Human Geography Map. Turn off the Human Geography Label reference layer.

3. On the Settings toolbar, in the Properties pane, under Appearance, adjust the Transparency slider to about 25%.

4. Under Blending, click Normal and change the selection to Multiply.

   This blend mode improves the look.

5. On the Settings toolbar, click Effects.

6. Turn on the Brightness & Contrast effect. Set the Brightness to **70** and the Contrast to **250**.

7. On the Contents pane, click Map Properties, and turn on the Preserve map scale toggle.

   Preserving the scale sets the map to open by default at the scale at which you saved it.

8. Zoom in or out so that the survey fills most of the screen.

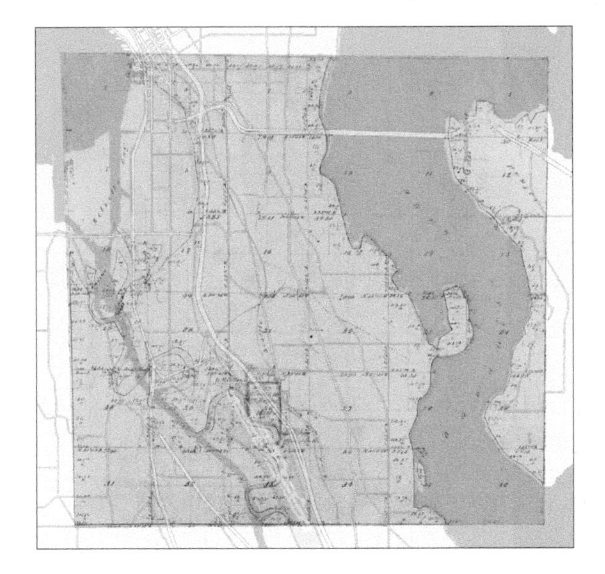

Save the map

1. On the Contents toolbar, click Save and open > Save as.
2. Save your map with the following settings:
   - Title: **Seattle Survey**
   - Folder: Top 20 Tutorial Content
   - Tags: **Top 20, survey**
   - Summary: **A historical survey of Seattle.**

   > *If you need to create the folder, click the down arrow and choose Create new folder. Type* **Top 20 Tutorial Content**.

3. Click Save.

Take the next step

Using a sketch layer, you can select a smaller portion of a media layer and apply effects to enhance the map. In the southwestern area, the change to the river is an interesting feature to highlight, so you'll crop to that extent and blend it into the basemap. You'll need the following skills: adding a sketch layer, adjusting layer visibility, and working with group layers.

Reference the image of the final map, and then follow these steps. Add a polygon sketch layer of the area surrounding the historical river. Set the fill to white and the transparency to zero percent, with no outline color. In the Layers pane, group the layers and order them so that the sketch layer is on top. With the sketch layer selected, in the Properties pane, change the blending to Destination Atop. In the Effects pane, apply the Bloom effect with these approximate settings: Strength 0.8 and Radius 1.5. With the Group layer selected, set the visibility range to start at Cities. Save your map.

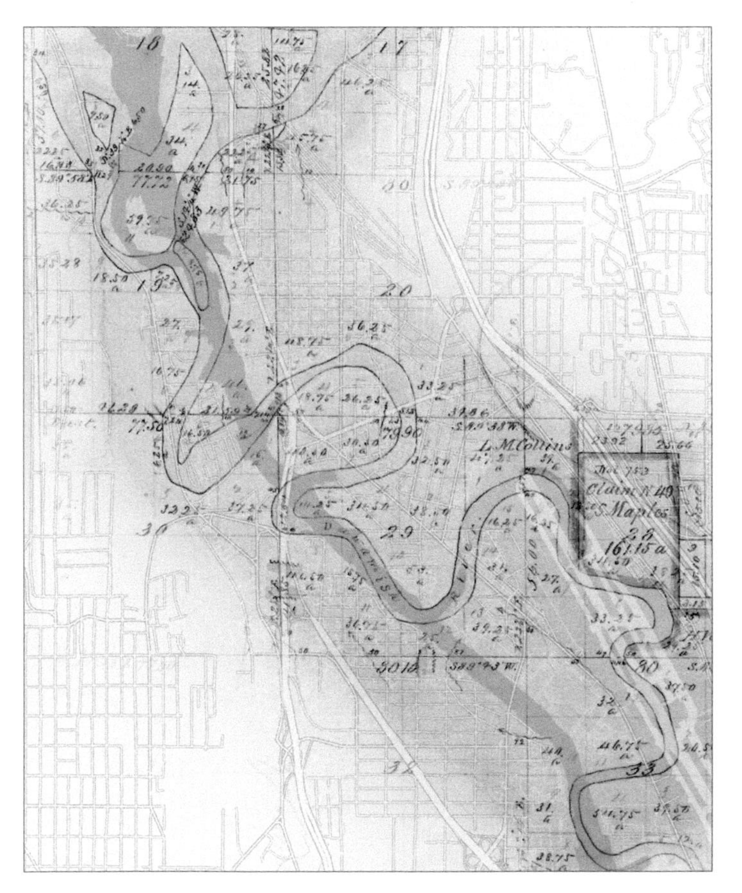

Summary

In this chapter, you learned how to add a media layer and position the layer using overlay mode. You then added a layer to a basemap and used it to position the media layer in side-by-side mode. Finally, you improved the appearance with simple visual enhancements.

Workflow

1. In the Add layers pane, click Add media layer and add an image.
2. Position the media layer in either overlay or side-by-side mode.
3. Blend or enhance the layer as needed.

CHAPTER 14

Creating forms

Objectives

- Use a form to edit attribute data.
- Learn how edits are applied to a table.
- Explore form configurations.
- Create a form.

Introduction

Forms are a customized data entry method for entering and updating attribute data. A configured form provides a step-by-step approach for editors to enter data for specific fields and defines the values that can be entered. Once you configure a form, you can use expressions to dynamically modify the form. As data is entered, the form can show new fields, hide unnecessary fields, or populate fields. Forms are a great way to streamline data entry and management and limit accidental errors.

In this chapter, you'll use a form and learn how the form was configured.

Tutorial 14-1: Explore a configured form

In this tutorial, you'll add a feature to a map and use a form to enter important attributes about a tree—work that a tree surveyor might do. By using a form to enter data, you'll see how the form functions.

Sign in and open a map

Open an existing map with a configured form.

1. Sign in to ArcGIS Online.
2. On the top navigation bar, click the magnifying glass to search for items in ArcGIS Online.
3. Type **owner:Top20EssentialSkillsForArcGISOnline Paris form** and press Enter.
4. Under Filters, turn off the filter for your organization and expand your search to all of ArcGIS Online.
5. On the Paris Tree Survey Form web map, click Open in Map Viewer.

   The map contains one layer, Tree Survey.

Add a feature and use a form

Forms simplify the process of adding attributes. To see how this works, you'll use a form.

> *Review the basics for navigating the user interface in Map Viewer. The Contents toolbar is the dark vertical toolbar on the left, and the Settings toolbar is the light vertical toolbar on the right. When you click or select something on a toolbar, a pane usually opens, allowing you to adjust a related setting. In the Settings toolbar, click Edit.*

1. On the Settings toolbar, click Edit.
2. In the Editor pane, under Create features, click New feature.

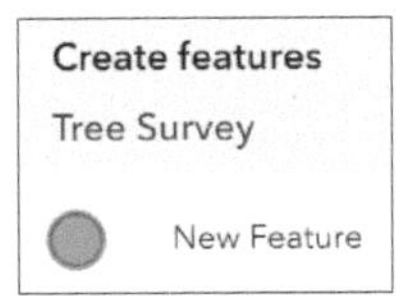

3. In the map, next to the Champs-Élysées, click to add a feature.

The Create features pane lets you enter attribute data based on the configured form. The first question is Tree Status.

4. Under Tree Status, click Excellent.

Tree Survey
Tree Status*
(●) Excellent
( ) Good
( ) Fair
( ) Poor
( ) Dead
Tree Condition
Record the tree condition
Common Name*
Enter the common name of the tree
Enter the common name of the tree

The next field, Common Name, has an asterisk next to it, indicating that it's required.

To see how the form can dynamically change based on inputs, you'll change the Tree Status from Excellent to Dead.

5. Under Tree Status, change the selection to Dead.

Tree Survey
Tree Status*
( ) Excellent
( ) Good
( ) Fair
( ) Poor
(●) Dead
Tree Condition
Record the tree condition
Common Name
Enter the common name of the tree
Enter the common name of the tree

Common Name no longer has an asterisk, indicating that it's no longer required.

If the tree is dead, it may not be possible to know what kind of tree it was, so the form has dynamically updated to ensure the best possible data integrity. If the type of tree can be determined, it can be entered, but if it's too hard to tell what kind of tree it was, the question is optional and won't require someone to enter possibly incorrect information.

6. Change the Tree Status back to Excellent.
7. In the Tree Condition group, for Common Name, type **Plane**.
8. For Height, type **26**.

   The Date Planted isn't editable and is provided only for reference.
9. Expand the Site Details group.
10. For Water, click Storm basin water.
11. Expand the Collection Details group.

    The Collection Date has automatically been set.
12. For Inspection Notes, type **Inspection completed**.

Tree Survey

Tree Status*

Excellent

Good

Fair

Poor

Dead

Tree Condition

Record the tree condition

Common Name*

Plane

Enter the common name of the tree

Height*

26

Date Planted

5/1/1990

Site Details

Describe the site

Water*

City water

Storm basin water

Collection Details

Record collection details

Collection Date

12/28/2023

01:52:50 PM

Inspection Notes

Inspection completed

20/256

13. At the bottom of the form, click Create.

    You've completed the form and added the necessary attributes. The edits you made have been saved.

View the updated table

You used a form to enter attributes, so next you'll view the updated table.

1. On the Contents toolbar, click Layers.
2. On the Tree Survey layer, click Options and click Show table.
3. Scroll through the table, review the fields, and note the values you added.

   It's easier to add attributes with the form than to click each of the cells in the table and edit it directly.

   > *Some attributes weren't included in the form, so some cells have missing values. There are also values included that you didn't enter, which is expected. In the second tutorial, you'll review the form and see why this is.*

Save and share your map

1. On the Contents toolbar, click Save and Open > Save as.
2. In the Save map window, add the following settings:
   - Title: **Paris Tree Survey <your initials>**
   - Folder: Top 20 Tutorial Content
   - Tags: **Top 20, form**
   - Summary: **A map showing a tree survey in Paris.**

   > *If you need to create the folder, click the down arrow and choose Create new folder. Type* **Top 20 Tutorial Content**.

3. Click Save.
4. On the Contents toolbar, click Share map.
5. In the Share pane, click Everyone (public) and click Save.

Tutorial 14-2: Review the configured form

Now you'll see how the form was configured. You'll click each element and view the properties.

Open the Configure form window

1. On the Settings toolbar, click Forms.

   The Configure form window appears. Form elements that are used are listed on the left.

Review the tree status

1. Click the Tree Status element.

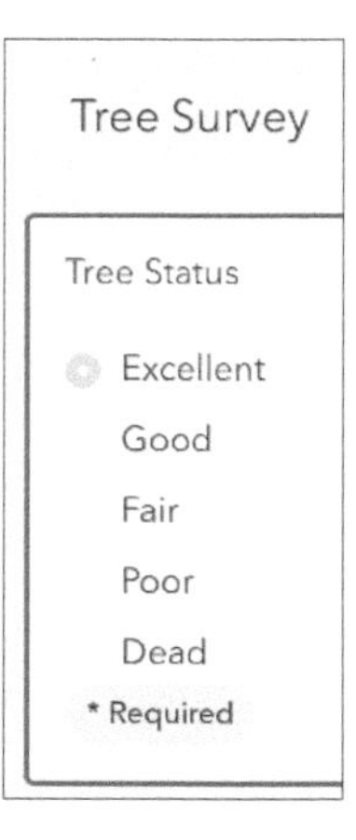

2. On the right, verify that the Properties pane has appeared.

   The Properties pane is divided into two parts: Formatting (top) and Logic (bottom).

3. In the Formatting section, under Input type, verify that the option for Radio buttons has been set.

   The radio buttons let you choose a value (for example, Excellent). These values were configured with the layer to limit the available options. You can describe a tree's status in many ways. By using a defined list of options, only five choices are possible, making analysis of the results easier.

   The Tree Status options come from a list.

Under Logic, Editable, Required, and Visible are checked for configuring the Tree Status element.

- [x] Editable
- [x] Required
- [x] Visible

Review the tree condition

This group of questions is related and organized into a group element. The ability to organize in this way is another advantage of digital forms.

1. Click the Tree Condition group element to select it.
2. In the Properties pane, confirm that the toggle is set to Expand initial state.

   The group element is set to be expanded when the form is opened, showing all the questions.

Review the common name

The first element in the Tree Condition group is Common Name.

1. In the group element, click the Common Name element.
2. In the Properties pane, under Logic, to the right of the Required check box, click the Settings button (gear).

   Field length*
   256
   Logic
   Set dynamic behavior with Arcade expressions
   Editable
   Required
   Live trees only
   expr
   Expressions
   Search expressions
   Live trees only
   expr
   1
   + New expression

   The Live trees only expression is shown. The checkmark indicates that it's being used.

3. Next to Live trees only, click the more button and click Edit.

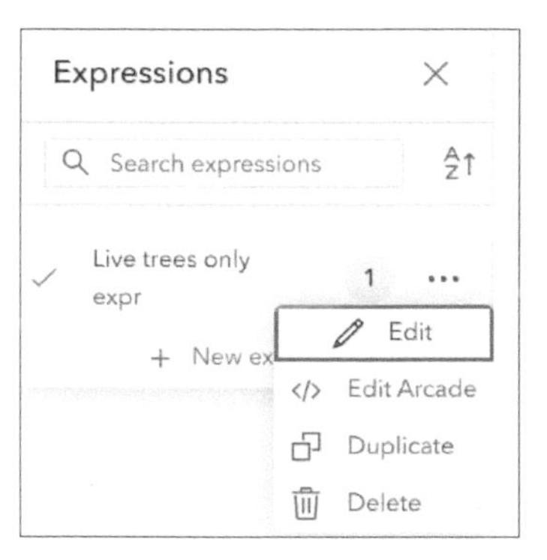

The Expression Builder window opens. The expression that must be satisfied for the Common Name question is as follows: Condition Is not Dead. As you learned, if the tree condition is any other option besides Dead, the question is required. If the condition is Dead, the Common Name question is no longer required.

4. Click Done.

In the Properties pane, in the Formatting section, text has been added under Placeholder. In the element, the placeholder text is shown in light gray, which provides a prompt when using the form.

Review the genus and species

When you reviewed the table, you may have seen the Genus and Species fields and noticed that they didn't have any values. These elements have been marked as hidden, so they didn't appear in the form and no data was entered for them.

1. Click the Genus element.

2. In the Properties pane, under Logic, confirm that the Visible check box is unchecked.

These elements have already been added to the form, so they can easily be made visible, but they are currently not needed.

Review the height

1. Click the Height element.

2. In the Properties pane, under Input type, confirm that the Input type is Number – Double.

   When you use this data type, the value entered can only be a number. Preventing the use of words for the height will make any future calculations from this field easier. Choosing an appropriate data type is a way to reduce unintentional errors. This field also uses placeholder text to explain that feet units should be used.

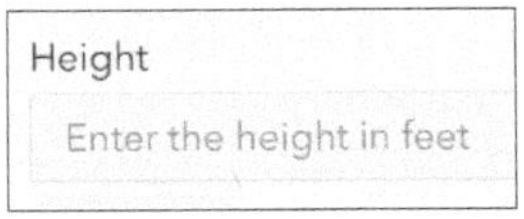

   Height
   Enter the height in feet

Review the date planted

1. Click the Date Planted element.

   Underneath the element, Read-only and Calculated have been selected. This survey is designed to collect data on the current status of trees that were planted many years ago. The planting date for all the trees is the same, so a calculated expression is used to automatically enter the same planting date.

2. In the Properties pane, under Logic, to the right of Calculated expression, click Settings.

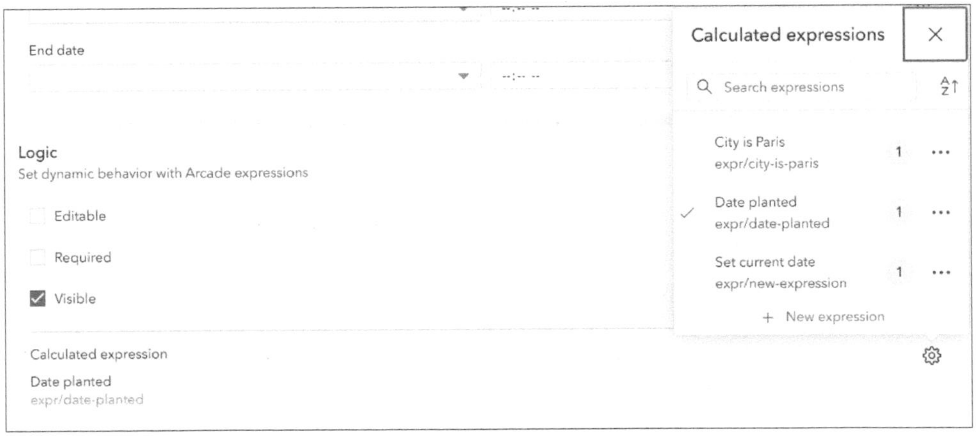

   End date
   Logic
   Set dynamic behavior with Arcade expressions
   Editable
   Required
   Visible
   Calculated expression
   Date planted
   expr/date-planted
   Calculated expressions
   Search expressions
   City is Paris
   expr/city-is-paris
   Date planted
   expr/date-planted
   Set current date
   expr/new-expression
   New expression

3. Next to Date planted, click the more button and click Edit Arcade.

   The Expression Builder shows a simple Arcade expression that adds a date.

4. Review the expression and close the builder.

5. In the Properties pane, under Logic, confirm that the Editable check box is unchecked.

   Since the expression will add the date value for this field, the person conducting the survey doesn't need to edit it. The Visible check box is checked, so it will still appear in the form because knowing the planting date may be of interest.

Review the city

A question for the City element didn't appear on the form, but a field for City did appear in the table, with the value "Paris" entered. Now you'll see how that happened.

1. Click the City element and view the Logic section.

   The City element isn't editable, required, or visible. Under Calculated expression, an expression is applied, making the field a calculated field.

On your own

As you did for the other expressions, in the Properties pane, review the expression used to calculate the City field.

All the trees being surveyed are in the same city, so the text "Paris" is calculated and applied to every feature, saving the need to repeatedly enter it. There is no reason to view this field when filling out the form, so it has been hidden to streamline the survey process.

Review the water

1. Click the Water element and view the Properties.

2. In the Formatting section, under Input type, confirm that the selected option is Combo box.

   Only two options are available: City water and Storm basin water. These options come from a list.

Review the collection date

In the Collection Details group element, the Collection Date element has been set to use a calculated expression.

1. Click the Collection Date element and view the Properties.

On your own

As you did for the other expressions, in the Properties pane, review the expression used to calculate the Collection Date field.

This expression uses an Arcade expression to automatically set the current date and time. No editing is needed, simplifying the surveying process.

2. Click Done.

Review the inspection notes

1. Click the Inspection Notes element and view the Properties.
2. Under Input type, confirm that the option has been set to Text – Multiline.

   This option lets users to add text. This element isn't required since users may not need to add notes for every tree.

Close the form

If you made accidental edits to the form, you can cancel and close the window.

1. In the Configure form window, click Cancel. Click OK.

   The window closes and Map Viewer remains visible.

Take the next step

Now that you've seen how a form works, configure a form using a new map and new layer.

Open a new map. Add a copy of the layer that was used in this chapter. To locate it, in the Add layer search box, type **owner:Top20EssentialSkillsForArcGISOnline essential form**. Add the layer. On the Settings toolbar, click Forms. In the Configure form window, drag fields from the right into the box on the left. Add fields you think are useful, select the elements, and configure them to function in a suitable way.

Create an expression to control the behavior of an element. When you are done, edit and add a feature to the map to test the form's functionality. Save your map with the name **Essential form <your initials>** in Top 20 Tutorial Content, and share the map publicly.

Summary

In this chapter, you used a form to enter attributes, reviewed a table, and verified the attribute updates. You then reviewed the form and saw how the elements were configured.

Workflow

1. Create a layer and add necessary fields.
2. Create lists for certain fields (optional).
3. Add the layer to a map.
4. Configure the form:
    - Configure elements.
    - Group related questions together.
    - Include expressions (optional).
5. Save the form and map.
6. Test the form.

CHAPTER 15

Creating a notebook

Objectives

- Explore the basics of ArcGIS Notebooks in ArcGIS Online.
- Learn how to create Python code in ArcGIS Notebooks.
- Create a web map, browse and add content, and check user profiles.
- Save and share a web map.

Introduction

ArcGIS Notebooks is designed to work with geospatial data. With Notebooks, you can write, document, and run Python code in your web browser. You can harness Python resources from Esri (ArcGIS API for Python and ArcPy™ to perform spatial analysis, map visualization, and other GIS tasks in the notebook environment. With the ArcGIS Python libraries, hundreds of popular third-party Python libraries are built in, such as Pandas, NumPy, and other AI and machine learning packages. This integration provides a seamless user experience for spatial analysis, data science, and content management.

You can run notebooks in ArcGIS Pro, ArcGIS Online, and ArcGIS Enterprise. This chapter concentrates on Notebooks in ArcGIS Online, teaching you how to build hosted notebooks in the cloud.

Tutorial 16-1 introduces you to notebooks. You'll learn the basics, including how to create a notebook in ArcGIS Notebooks, work with content cells in the notebook, and complete a GIS workflow for creating a map. You'll check a user profile, browse for a layer, and save a map to ArcGIS Online using a notebook and Python API.

Tutorial 15-1: Explore the basics of ArcGIS Notebooks

The notebook runtime determines resource allocations, such as memory, CPU, and the availability of Python packages. Standard runtime contains standard open-source Python libraries and Python API, which are all you'll for your first notebook.

> *Advanced runtime contains the Standard runtime libraries, plus ArcPy and increased memory and CPU. Advanced lets you manage larger analyses and processing tasks. Advanced also has a GPU support option for workflows involving machine learning and deep learning.*

In this tutorial, you'll use Standard runtime.

Choose an option for completing this tutorial

You have two options for completing this tutorial. We recommend that you try option 1, which will give you the experience of creating your own notebook and entering code.

- Option 1: Open a new notebook, follow the tutorial steps, and complete the work in your notebook.
- Option 2: Open a prepared notebook, read along, and run the code cells.

These instructions apply to option 1. If you prefer using option 2, the instructions that appear in the notebook will be slightly modified.

Option 1

1. Sign in to ArcGIS Online.

2. On the top navigation bar, click Notebook.

3. Click New Notebook.

   The list has three options: Standard, Advanced, and Advanced with GPU support.

4. Click Standard.

5. Jump to the "Save Your Notebook" section to complete the tutorial.

Option 2

1. In a browser, go to **links.esri.com/Top20AGOLNotebook.**
2. Sign in.
3. On the Getting Started with ArcGIS Online Notebooks item page, click Open notebook.
4. Complete the rest of this tutorial in the notebook.

Save your notebook

Naming and saving your notebook as soon as possible after you create it is important.

1. In the upper right of the notebook, click Save and click Save As.
2. Add the following details:
   - Title: **My Notebook <your initials>**
   - Tags: **Top 20, notebook**
   - Summary: **My first notebook.**
   - Folder: Top 20 Tutorial Content

   > *If you haven't created this folder, save the notebook in your root folder.*

3. Click Save Notebook.

   This information will show up on the item page for your notebook in ArcGIS Online.

> **Hint:** It's a good idea to save your work regularly when you work in notebooks that are hosted online. If there is no Python activity in the notebook for 30 minutes, the Python kernel will shut down, the notebook will stop working, and all variables in memory will be lost. Once you've restarted the kernel, you would need to run all the cells again from the beginning to restore the values.

Work with a notebook

Each new notebook starts with several markdown cells that are already populated and one code cell that calls Python API to connect you to ArcGIS Online. After that, you can add code and markdown cells to create your workflow.

1. Double-click the Welcome to your notebook cell to make it editable.

```
## Welcome to your notebook.
```

   This is a markdown cell. Markdown is a lightweight, plain-text formatting syntax that's widely used on the internet. After you double-click the cell, the text appears in blue with two number signs (##) in front of it.

2. Click Run.

   The cell runs and turns into a header.

> **Hint:** The keyboard shortcut for running cells in a notebook is Shift + Enter. To see all shortcuts, click the command palette at the top of the notebook.

3. Double-click the Welcome to your notebook cell again to add two more number signs (##) in the cell.

   You should now see four number signs.

4. Run the cell.

   The additional number signs decrease the size of the header.

5. Double-click the Welcome to your notebook cell again. Remove two of the number signs from the header. At the end of the line, press Enter to add a new line.

6. On the second line, type **This is my first notebook.**

```
## Welcome to your notebook.
This is my first notebook
```

7. Click Run.

   The header returns to its initial size and the text you entered is visible.

In a new notebook, the second cell will say "Run this cell to connect to your GIS and get started."

This cell directs you to run the code cell that follows it, which connects you to ArcGIS Online. You must run the cell. If you don't, the other code cells won't run.

8. Click inside the code cell to select it.

```
from arcgis.gis import GIS
gis = GIS("home")
```

9. Click Run.

While the cell runs, to the left of the cell, an asterisk appears inside the brackets in the input area, so that it appears as `[*]`.

After the cell finishes running, the number 1 replaces the asterisk in the brackets so that it appears as `[1]`. The number in the brackets increases by one each time a code cell is run.

> **Hint:** If you make a mistake and want to rerun cells, the number in the brackets will increase and may thus appear different from what your colleagues see as they run cells. This number doesn't affect your code, however.

Check a user profile in ArcGIS Notebooks

First, you'll check the profile for your own account.

With an `arcgis.gis.User` object, you can retrieve biographical information about other members. If no biographical information or profile photo is set, these properties will return `None`.

1. In the blank cell, type the following, using the image as a guide:

```
me = gis.users.me
theuser = me.username
me
```

```
me = gis.users.me
theuser = me.username
me
```

2. Run the cell.

   Your ArcGIS Online profile appears.

Convert a code cell into a markdown cell

Next, you'll convert the new code cell that automatically appeared into a markdown cell and create a heading.

1. Click inside the new code cell.
2. In the menu, click Cell, point to Cell Type, and click Markdown.

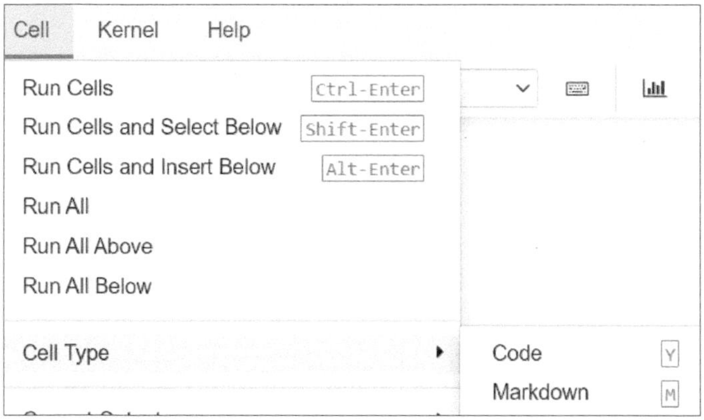

   Cell Kernel Help
   Run Cells Ctrl-Enter
   Run Cells and Select Below Shift-Enter
   Run Cells and Insert Below Alt-Enter
   Run All
   Run All Above
   Run All Below
   Cell Type
   Code Y
   Markdown M

   A new markdown cell is created.

3. In the cell, type **### My Map**. Press Enter.
4. In the same cell, on the second line, type **Yellowstone, WY**.

   ```
   ### My Map
   Yellowstone, WY
   ```

5. Run the cell.

Create a map in ArcGIS Notebooks

Now that you have added a heading, you're ready to add a map.

In the new code cell, you'll create a variable named my_map that represents the map and use Python API to set the variable to a web map that is centered over Yellowstone National Park, Wyoming.

1. In the new blank code cell, enter the following code:
   `my_map = gis.map("Yellowstone National Park, WY").`

   ```
   my_map = gis.map("Yellowstone National Park, WY")
   ```

2. Run the cell.

   Now that you've named your variable, you'll call it to create the map.

3. In the blank cell, type **my_map**.
4. Run the cell.

   After the cell runs, your first map will appear in your notebook.

On your own

You can change the location of the center of the map by going back to the previous cell, changing "Yellowstone National Park, WY" to a different location, and running the cells again.

5. Pan and zoom around the map to explore it.

Tutorial 15-2: Explore ArcGIS Online content in a notebook

In Notebooks, you can search for content in the cloud and view item metadata. Next, you'll search for a layer in ArcGIS Living Atlas and add it to the notebook.

1. On the top ribbon of the notebook editor, click Add.

The content browser panel lets you add an ArcGIS Online item directly to the notebook. You can access your own content, content shared with you, or public content from ArcGIS Online or ArcGIS Living Atlas.

2. Click My content and click Living Atlas. In the search field, type **recent earthquakes**.

3. Find the Recent Earthquakes layer owned by Esri. On the layer, click Add to add it to your notebook.

   A new code cell containing a code snippet is added to the notebook below your map. This cell calls the layer as the variable and loads its metadata.

```
# Item Added From Toolbar
# Title: Recent Earthquakes | Type: Feature Service | Owner: esri_livefeeds2
item = gis.content.get("9e2f2b544c954fda9cd13b7f3e6eebce")
item
```

4. Run the code cell.

   The item metadata is added as an object to the notebook below the cell, but it's not yet added to the map. You'll learn some shortcuts for writing code and displaying relevant documentation. In the blank code cell, you'll call your map variable.

5. Type **my_** and press Tab on your keyboard.

   Pressing Tab while typing a variable name activates the notebook's autocomplete functionality. The line completes and displays `my_map`.

6. After `my map`, type **.ad** so it appears as `my_map.ad`.

7. Press Tab.

   Options for completing the code are shown.

8. Click the `add_layer` option.

> **Hint:** Map view objects have a method named `add_layer`. This method lets you add a layer to a map in ArcGIS Notebooks.

The completed code is `my_map.add_layer`.

9. Press Shift + Tab.

   Pressing Shift + Tab opens the docstring, a small piece of documentation for developers that describes what the method does.

10. After investigating the docstring, click the X to close it.

    You can access the help documentation by typing a question mark at the end and running the cell.

11. At the end of your code, add a question mark so that your code says `my_map.add_layer?` and run the cell.

    After reading the help documentation, close it.

12. In the cell, delete the question mark you added.

Add a layer to the map

In ArcGIS Notebooks, you can add layers to maps and build map view objects that can be saved as a web map. Next, you'll add the Recent Earthquakes layer to the map of Yellowstone National Park.

1. In the same cell, type **`(item)`**.

   ```
   my_map.add_layer(item)
   ```

2. Run the cell.

3. After the cell runs, pan and zoom around the map to confirm that the Recent Earthquakes layer has been added to the map.

   You've created your own map and added a layer to it using ArcGIS Notebooks. The map includes zoom buttons, a compass, and the option to change from a 2D map view to a global scene view. This web map has the same functionality as the maps you use in ArcGIS Online.

Save the map to ArcGIS Online

With a few lines of code, you can save your map to ArcGIS Online. Web maps are defined by specific properties, such as title, description (snippet), and tags. You can define these properties by creating a dictionary that contains them in Python. Then you can save the map as a web map.

1. In a new code cell, type the following code:

```
webmap_properties = {'title':'My Map', 'snippet': 'My map from my
notebook', 'tags':['ArcGIS Notebooks', 'Yellowstone National Park',
'Recent Earthquakes']}
```

2. Run the cell.

3. In a new code cell, add the following: `my_map.save(webmap_properties)`.

4. Run the cell.

   Running this cell creates an active link that will take you to the web map in ArcGIS Online.

5. In the cell, click the active link and verify that the web map was created in ArcGIS Online.

6. Close the browser tab with the web map item and return to your notebook.

7. In the upper right of your notebook, click Save > Save.

Share a notebook

You can share a notebook with other users in your organization or share it publicly. If other users have access to ArcGIS Notebooks, they can open and run the notebook themselves. Opening a copy of a notebook creates a copy of it in the ArcGIS Online content of the user opening it. A user who isn't authorized to run ArcGIS Notebooks can only download a notebook. When you download a notebook, you can open it in ArcGIS Pro or Jupyter.

1. On the top ribbon, click Share.

   A panel opens on the left side of the notebook.

2. In the Share panel, click Share.

   You can share the notebook with your organization, publicly, or with a group.

3. Click Everyone (public) and click Save.

4. In the upper left of the notebook, click the menu button and click Content.

   Your notebook and web map are available for review in Content.

Take the next step

Now that you've learned how to use Python and ArcGIS Notebooks, create a new notebook, search for content, add multiple layers, and publish a web map to ArcGIS Online.

Summary

In this chapter, you created a notebook in ArcGIS Notebooks. You entered markdown and code content and reviewed your user profile. You completed a common GIS workflow for creating a map, adding a layer to it, and publishing it to ArcGIS Online using ArcGIS Notebooks and Python API.

CHAPTER 16
Preparing data with a notebook

Objectives

- Learn how to perform analysis with ArcGIS Notebooks in ArcGIS Online.
- Integrate open-source Python libraries to perform data engineering tasks.
- Schedule a notebook task in ArcGIS Online.

Introduction

This chapter builds on the basics of ArcGIS Notebooks, covered in chapter 15, and provides a more advanced workflow for getting data ready for spatial analysis and tools. Notebooks use an interactive environment in which you can code, execute, and visualize your analysis steps in real time. While using the library of geospatial data and analysis tools, you can document and share your steps, providing other users with a reproducible workflow.

Tutorial 16-1: Prepare data for spatial analysis

You'll use ArcGIS Notebooks and some third-party open-source Python libraries to handle data engineering tasks such as renaming field names, detecting and filling missing values, correcting data types, and creating pivot tables. You'll also learn how to geoenable the data by joining it to an existing boundary layer available in ArcGIS Living Atlas, using ArcGIS API for Python. Finally, you'll publish a map, showing your results.

> *The voter turnout dataset used in this tutorial is obtained from the Harvard Dataverse and the US Census Bureau. The dataset has vote totals from each US county for presidential elections from 2000 to 2020.*

Choose an option for completing this tutorial

You have two options for completing this tutorial. We recommend that you try option 1, which will give you the experience of creating your own notebook and entering code.

- Option 1: Open a new notebook, follow the tutorial steps, and complete the work in your notebook.
- Option 2: Open a prepared notebook, read along, and run the code cells.

The following instructions apply to option 1. If you prefer using option 2, the instructions that appear in the notebook will be slightly modified.

Option 1

1. Sign in to ArcGIS Online.
2. On the top navigation bar, click Notebook.
3. Click New Notebook.

   The list has three options: Standard, Advanced, and Advanced with GPU support.
4. Click Standard.
5. Go to the "Save Your Notebook" section to complete the tutorial.

Option 2

1. In a browser, go to **links.esri.com/Top20AGOLNotebook2**.
2. Sign in.

3. On the Prepare Data for Spatial Analysis item page, click Open notebook.

4. Complete the rest of this tutorial in the notebook.

Save your notebook

> **Hint:** It's a good idea to save your work regularly when you work in notebooks that are hosted online. If there is no Python activity in the notebook for 20 minutes, the Python kernel will shut down, the notebook will stop working, and all variables in memory will be lost. Once you've restarted the kernel, you would need to run all the cells again from the beginning to restore the values.

1. In the upper right of the notebook, click Save and click Save As.

Save
Save As

2. Add the following details:

   - Title: **My Data Engineering Notebook <your initials>**
   - Tags: **Top 20, engineering, notebook**
   - Summary: **My data engineering notebook.**
   - Folder: Top 20 Tutorial Content

> *If you haven't created this folder, save the notebook in your root folder.*

3. Click Save Notebook.

4. To connect to your GIS, in the notebook, click inside the prepopulated cell. On the toolbar, click Run.

```
from arcgis.gis import GIS
gis = GIS("home")
```

> **Hint:** To run a cell, you can also press Shift + Enter.

Import open-source Python libraries

ArcGIS Notebooks integrates hundreds of popular third-party Python libraries, such as Pandas, NumPy, and other AI and machine-learning packages. Pandas provides functions for handling data engineering tasks such as detecting and filling missing values, identifying and removing outliers, correcting data types, and formatting strings. Integrating Python libraries within ArcGIS Notebooks lets you apply advanced data cleansing and use tools and algorithms from the Python ecosystem. Next, you'll import the necessary open-source Python modules.

1. In the blank code cell, type the following:

```
import pandas as pd
import os
```


```
import pandas as pd
import os
```


2. Run the cell.

Load and prepare election data using Pandas

Next, you'll search for the voter turnout dataset (a CSV file) shared in ArcGIS Online and add it to the notebook.

1. On the notebook editor ribbon, click Add.

The content browser lets you add an ArcGIS Online item directly to the notebook. You can access your own content, content shared with you, or public content from ArcGIS Online or ArcGIS Living Atlas. You'll add content from ArcGIS Online.

2. Click My content and choose ArcGIS Online.
3. In the search box, type **owner:Top20EssentialSkillsForArcGISOnline countypres**.
4. On the countypres_2000_2020 item, click Add.
5. Close the content browser panel.

A new code cell containing a code snippet is added to the notebook. This cell calls the layer as the variable and loads its metadata.

```
# Item Added From Toolbar
# Title: countypres_2000_2020 | Type: CSV | Owner: Top20EssentialSkillsForArcGISOnline
item = gis.content.get("cdd47d75cc6d4dcd8e75e42de370881d")
item
```

6. Run the code cell.

   You'll download the CSV file and use the Pandas read function to load the county election dataset into the data frame. To prevent inadvertently deleting the leading 0 character used by some counties, you must ensure that the county_fips field is read as an object and the data type used is not a number.

7. Enter and run the following code:

   ```
   elections_data_path = item.download()
   ```

   Next, you'll use Pandas functionality to create a data frame. A Pandas data frame is a tabular data structure of columns and rows. The columns are referred to as *attributes*, or *attribute fields*, and the rows are referred to as *records*. To create a data frame, you first define a variable for the dataset.

8. Enter and run the following code:

   ```
   elections_complete_df = pd.read_csv(elections_data_path, dtype =
   {"county_fips":object})
   ```

   Next, you'll confirm that the dataset loaded properly before you go on.

9. Enter and run the following code:

   ```
   elections_complete_df
   ```

   A portion of the table is shown.

	year	state	state_po	county_name	county_fips	office	candidate	party	candidatevotes	totalvotes	version	mode
0	2000	ALABAMA	AL	AUTAUGA	1001	US PRESIDENT	AL GORE	DEMOCRAT	4942	17208	20220315	TOTAL
1	2000	ALABAMA	AL	AUTAUGA	1001	US PRESIDENT	GEORGE W. BUSH	REPUBLICAN	11993	17208	20220315	TOTAL
2	2000	ALABAMA	AL	AUTAUGA	1001	US PRESIDENT	RALPH NADER	GREEN	160	17208	20220315	TOTAL
3	2000	ALABAMA	AL	AUTAUGA	1001	US PRESIDENT	OTHER	OTHER	113	17208	20220315	TOTAL
4	2004	ALABAMA	AL	AUTAUGA	1001	US PRESIDENT	JOHN KERRY	DEMOCRAT	4758	20081	20220315	TOTAL
...	...	...	...	...	...	...	...	...	...	...	...	...
72560	2016	WYOMING	WY	WESTON	56045	US PRESIDENT	OTHER	OTHER	194	3526	20220315	TOTAL
72561	2020	WYOMING	WY	WESTON	56045	US PRESIDENT	JOSEPH R BIDEN JR	DEMOCRAT	360	3560	20220315	TOTAL
72562	2020	WYOMING	WY	WESTON	56045	US PRESIDENT	JO JORGENSEN	LIBERTARIAN	46	3560	20220315	TOTAL
72563	2020	WYOMING	WY	WESTON	56045	US PRESIDENT	OTHER	OTHER	47	3560	20220315	TOTAL
72564	2020	WYOMING	WY	WESTON	56045	US PRESIDENT	DONALD J TRUMP	REPUBLICAN	3107	3560	20220315	TOTAL

Some of the table headings should be improved to better identify the data in the column. First, you'll rename these headings and then use the Pandas head function to return the first five rows of the table to confirm that the new headings are correct.

10. Enter and run the following code:

```
rename_cols = {
    "state_po": "state_abbr",
    "county_fips": "FIPS",
    "party": "pol_identity"
}
elections_complete_df.rename(columns=rename_cols, inplace=True)
elections_complete_df.head()
```

The result is shown.

	year	state	state_abbr	county_name	FIPS	office	candidate	pol_identity	candidatevotes	totalvotes	version	mode
0	2000	ALABAMA	AL	AUTAUGA	1001	US PRESIDENT	AL GORE	DEMOCRAT	4942	17208	20220315	TOTAL
1	2000	ALABAMA	AL	AUTAUGA	1001	US PRESIDENT	GEORGE W. BUSH	REPUBLICAN	11993	17208	20220315	TOTAL
2	2000	ALABAMA	AL	AUTAUGA	1001	US PRESIDENT	RALPH NADER	GREEN	160	17208	20220315	TOTAL
3	2000	ALABAMA	AL	AUTAUGA	1001	US PRESIDENT	OTHER	OTHER	113	17208	20220315	TOTAL
4	2004	ALABAMA	AL	AUTAUGA	1001	US PRESIDENT	JOHN KERRY	DEMOCRAT	4758	20081	20220315	TOTAL

The election data includes records that are missing data in the FIPS field. This missing data is referred to as *null values.* You'll join this table to an ArcGIS Living Atlas US counties boundary layer based on the FIPS value, so this value cannot be null.

Next, you'll identify how many rows have null values and establish a strategy for dealing with these null values.

11. Enter and run the following code:

```
elections_complete_df.query("FIPS.isnull()")
```

The first five results are shown.

	year	state	state_abbr	county_name	FIPS	office	candidate	pol_identity	candidatevotes	totalvotes	version	mode
8085	2000	DISTRICT OF COLUMBIA	DC	DISTRICT OF COLUMBIA	NaN	US PRESIDENT	AL GORE	DEMOCRAT	171923	201894	20220315	TOTAL
8086	2000	DISTRICT OF COLUMBIA	DC	DISTRICT OF COLUMBIA	NaN	US PRESIDENT	GEORGE W. BUSH	REPUBLICAN	18073	201894	20220315	TOTAL
8087	2000	DISTRICT OF COLUMBIA	DC	DISTRICT OF COLUMBIA	NaN	US PRESIDENT	RALPH NADER	GREEN	10576	201894	20220315	TOTAL
8088	2000	DISTRICT OF COLUMBIA	DC	DISTRICT OF COLUMBIA	NaN	US PRESIDENT	OTHER	OTHER	1322	201894	20220315	TOTAL
8089	2004	DISTRICT OF COLUMBIA	DC	DISTRICT OF COLUMBIA	NaN	US PRESIDENT	JOHN KERRY	DEMOCRAT	202970	227586	20220315	TOTAL

These rows represent votes cast in Washington, DC, which have null values (listed as NaN) for the FIPS field. Without these records, hundreds of thousands of votes wouldn't be counted in the analysis for 2020. You'll add the correct FIPS code for Washington, DC: 11001.

12. Enter and run the following code:

```
elections_complete_df.loc[elections_complete_df['state_abbr'] ==
 – 'DC', 'FIPS'] = '11001'
```

> *Code that should be entered as one continuous line is indicated with an em-dash highlighted in yellow. The dash should not be typed as part of the code.*

On your own

Rerun the code to identify records with null values:

```
elections_complete_df.query("FIPS.isnull()")
```

No records should be returned.

For this analysis, you'll focus on the 2020 election only.

13. Enter and run the following code:

```
elections_df_2020 = elections_complete_df.query("year == 2020")
elections_df_2020.head()
```

	year	state	state_abbr	county_name	FIPS	office	candidate	pol_identity	candidatevotes	totalvotes	version	mode
16	2020	ALABAMA	AL	AUTAUGA	1001	US PRESIDENT	JOSEPH R BIDEN JR	DEMOCRAT	7503	27770	20220315	TOTAL
17	2020	ALABAMA	AL	AUTAUGA	1001	US PRESIDENT	OTHER	OTHER	429	27770	20220315	TOTAL
18	2020	ALABAMA	AL	AUTAUGA	1001	US PRESIDENT	DONALD J TRUMP	REPUBLICAN	19838	27770	20220315	TOTAL
35	2020	ALABAMA	AL	BALDWIN	1003	US PRESIDENT	JOSEPH R BIDEN JR	DEMOCRAT	24578	109679	20220315	TOTAL
36	2020	ALABAMA	AL	BALDWIN	1003	US PRESIDENT	OTHER	OTHER	1557	109679	20220315	TOTAL

Restructure the table

Even after correcting for the missing FIPS values, the format of the election data table prevents a proper join to the ArcGIS Living Atlas county boundary layer because each record corresponds to a candidate's votes for each county in each vote mode. You must reformat the table so that each record corresponds to a single county for each election year, with fields showing the total votes for each major-party candidate for that election year in that county. You can do this using the Pivot Table geoprocessing tool or Excel pivot tables, but you'll use Python.

First, you'll limit the records to candidates who identify themselves as either a Democrat or a Republican.

1. Enter and run the following code:

```
elections_df_party = elections_df_2020.query("pol_identity in
 ➥ ['DEMOCRAT', 'REPUBLICAN']")
```

Next, you'll find the total number of votes for the Republican and Democratic candidates in the 2020 election for each county in every state. In other words, you'll summarize the candidatevotes field, grouped by county, state, and political party.

2. Enter and run the following code:

```
candidate_votes = elections_df_party.groupby(['FIPS',
 ➥ 'county_name', 'state', 'pol_identity','year'])
 ➥ ['candidatevotes'].sum()

candidate_votes.head()
```

Next, you'll remove the multi-index created by the group by and sum, returning a new data frame.

3. Enter and run the following code:

```
candidate_votes_df = candidate_votes.reset_index()
candidate_votes_df.head()
```

	FIPS	county_name	state	pol_identity	year	candidatevotes
0	10001	KENT	DELAWARE	DEMOCRAT	2020	44552
1	10001	KENT	DELAWARE	REPUBLICAN	2020	41009
2	10003	NEW CASTLE	DELAWARE	DEMOCRAT	2020	195034
3	10003	NEW CASTLE	DELAWARE	REPUBLICAN	2020	88364
4	10005	SUSSEX	DELAWARE	DEMOCRAT	2020	56682

The previous table contains two records for each US county. One record shows the county's votes for the Republican party, and the other record shows votes for the Democratic party. You need a table with one record for each US county, with columns showing the total number of votes for the Republican or Democratic party. You'll create a pivot table to rearrange the rows and columns.

Pivot the data frame

The index "locks" the fields that stay the same for each county. The columns determine which values become new fields; in this case, the two values in the pol_identity column become two new fields. The values column determines which values are reported for each of the new fields.

1. Enter and run the following code:

```
elections_pivot_df = candidate_votes_df.pivot(
    index=[ 'FIPS', 'county_name', 'state', 'year'],
    columns=['pol_identity'],
    values=['candidatevotes']
)
```

2. To see the result, enter and run the following code:

```
elections_pivot_df.head()
```

				candidatevotes	
			pol_identity	DEMOCRAT	REPUBLICAN
FIPS	county_name	state	year		
10001	KENT	DELAWARE	2020	44552	41009
10003	NEW CASTLE	DELAWARE	2020	195034	88364
10005	SUSSEX	DELAWARE	2020	56682	71230
1001	AUTAUGA	ALABAMA	2020	7503	19838
1003	BALDWIN	ALABAMA	2020	24578	83544

Next, you'll remove the multi-index since you no longer need these fields to be locked for the pivot.

3. Enter and run the following code:

```
elections_pivot_df.columns =
 - elections_pivot_df.columns.get_level_values(1).rename(None)

elections_pivot_df.reset_index(inplace=True)
```

Now you'll rename the columns to better reflect their new meaning, check the output, and create a new data frame.

4. Enter and run the following code:

```
elections_pivot_df_new =
 ➥ elections_pivot_df.rename(columns={"DEMOCRAT": "votes_dem",
 ➥ "REPUBLICAN": "votes_gop"},)
```

You want to check the new data frame.

5. Enter and run the following code:

```
elections_pivot_df_new.head()
```

	FIPS	county_name	state	year	votes_dem	votes_gop
0	10001	KENT	DELAWARE	2020	44552	41009
1	10003	NEW CASTLE	DELAWARE	2020	195034	88364
2	10005	SUSSEX	DELAWARE	2020	56682	71230
3	1001	AUTAUGA	ALABAMA	2020	7503	19838
4	1003	BALDWIN	ALABAMA	2020	24578	83544

Add and publish the CSV file to ArcGIS Online

Now that the data is cleaned up, you'll export your Pandas data frame to a CSV file using the `to_csv` method. If you make a mistake, the code uses an `if/else` conditional statement to verify whether the CSV file already exists. The statement removes a pre-existing file and lets you publish a corrected file.

> *In these sections, you must modify the code before running it so that the items you're creating are unique within your ArcGIS organization. To make the items unique, you must add your initials. If you encounter an error, it may be because someone else in your organization has completed this tutorial using the item name that you're trying to use. Add unique characters to make your items unique.*

1. Enter the following code, but do not run it:

```
csv_file_path = '/arcgis/home/elections2020table.csv'
 ➥ if os.path.exists(csv_file_path):
    os.remove(csv_file_path)
    elections2020csv = elections_pivot_df_new.to_csv
 ➥ (csv_file_path)

else:
    elections2020csv = elections_pivot_df_new.to_csv
 ➥ (csv_file_path)
```

Before running this code, you must modify the item you're creating to make it unique within your organization.

2. Add your initials to the first line to make the name of the CSV file that you're creating unique—for example, modify the first line as follows: `csv_file_path = '/arcgis/home/elections2020tableABC.csv'`. Replace "ABC" with your initials.

3. Run the modified code.

   You'll upload the CSV file to ArcGIS Online using Python API. If the item already exists, it will be removed first.

4. Enter the following code, but do not run it:

```
csv_properties = {
    "title": "Election 2020 Voter Turnout",
    "tags": "csv, election, Top20EssentialSkillsForArcGISOnline",
    "type": "CSV"
    }
CSVs_to_delete = gis.content.search(query='title:Election 2020
➥ Voter Turnout', item_type="CSV")

if CSVs_to_delete:
    csv_to_delete = CSVs_to_delete[0]
    csv_to_delete.delete()
    Election2020CsvItem = gis.content.add(csv_properties,
➥ csv_file_path)

else:
    Election2020CsvItem = gis.content.add(csv_properties,
➥ csv_file_path)
```

   Before running this code, you'll modify the item to make it unique within your organization. You'll add your initials to the second and sixth lines.

5. Modify the second line as follows: `"title": "Election 2020 Voter Turnout ABC",`.

6. Modify the sixth line as follows: `CSVs_to_delete = gis.content.search( query='title:Election 2020 Voter Turnout ABC', item_type="CSV")`. In steps 5 and 6, replace "ABC" with your initials.

7. Run the modified code.

   Next, you'll publish the CSV item as a hosted layer. If the item already exists, it will be removed first.

8. Enter the following code, but do not run it:

```
features_to_delete = gis.content.search(query='title:Election 2020
 - Voter Turnout', item_type="Feature Layer")
if features_to_delete:
    feature_to_delete = features_to_delete[0]
    feature_to_delete.delete()
    Election2020CsvHosted = Election2020CsvItem.publish()
else:
    Election2020CsvHosted = Election2020CsvItem.publish()
```

   Before running this code, you'll modify the item to make it unique within your organization. You'll add your initials to the first line.

9. Add your initials as follows: `features_to_delete = gis.content.search (query='title:Election 2020 Voter Turnout ABC', item_type="Feature Layer")`. Replace "ABC" with your initials.

10. Run the modified code.

Run the Join Features Tool to geoenable the data

To use this data for spatial analysis, you need to include location information to determine where each county is located on a map. This section shows how to geocode the data by joining it to existing county geometries.

You can use resources such as ArcGIS Living Atlas to find geoenabled data. Each record in your election data represents information for a county, so you'll use an ArcGIS Living Atlas dataset that represents county geometry. You'll search for a US county boundary layer from ArcGIS Living Atlas and add it to the notebook.

1. On the notebook editor ribbon, click Add.

2. In the content browser, click My content and click Living Atlas. In the search box, type **USA Census Counties.**

3. Locate the USA Census Counties layer owned by Esri. On the layer, click Add to add it to your notebook.

4. Close the content browser.

5. Run the cell.

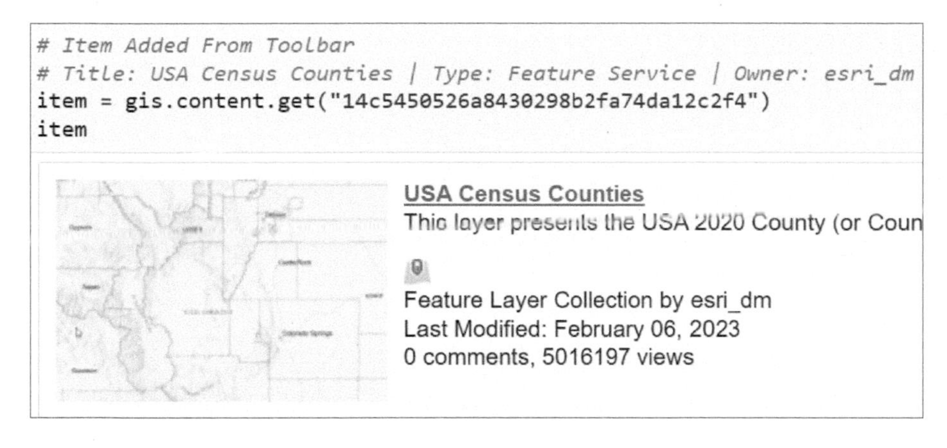

Next, you'll use the Join Features tool (analysis) to join the election 2020 data table to the USA Census Counties layer. If the resulting feature layer already exists, you'll remove it first.

> *This analysis step will consume about six credits.*

6. Enter the following code, but do not run it:

```
oldresults_to_delete = gis.content.search(query='title:
 ➥ Election2020VoterResultsByCounty', item_type="Feature Layer")
if oldresults_to_delete:
    oldresult_to_delete = oldresults_to_delete[0]
    oldresult_to_delete.delete()
```

   Before running this code, you'll add your initials to the first line.

7. Add your initials as follows: `oldresults_to_delete = gis.content.search(query='title:Election2020VoterResultsByCounty` **`ABC`**`', item_type="Feature Layer")`. Replace "ABC" with your initials.

8. Run the modified code.

   The ArcGIS Notebooks editor provides an Analysis pane that you can use to browse or add the code snippets for the analysis tools available in Map Viewer. You'll search for the Join Features tool in the Analysis pane and add it to the notebook.

9. On the ribbon, click the Analysis button.

10. In the Analysis tools pane, expand Summarize Data. To the right of Join Features, click the Insert into notebook button.

A new code block is added to your notebook. This code imports the `features` module from Python API and calls the `join_features()` function.

Before running the code, you'll review the function signature to see the parameters required for the `join_features` function. You must type each parameter exactly as the function expects, or an error will occur.

11. Replace the open and closed parentheses `()` with a question mark (**?**) and run the cell.

```
from arcgis import features
features.summarize_data.join_features?
```

This explains all the parameters that the `join_features` function uses as inputs. Some of these are optional.

12. Close the function explanation.

Next, you'll populate the function with the appropriate input parameters to run the analysis.

13. On the second line, type **`results`** = before `features.summarize`.

This change sets a new variable that will hold the result of the analysis process and names it Election2020VoterResultsByCounty.

14. Continue to edit the code as shown, but do not run the cell.

```
from arcgis import features
results = features.summarize_data.join_features(
    target_layer = item,
    join_layer = Election2020CsvHosted,
    join_operation ="joinOneToOne",
    attribute_relationship=[
        {
            "targetField": "FIPS",
            "operator": "equal",
            "joinField": "FIPS",
        }
    ],
    output_name="Election2020VoterResultsByCounty",
)
```

Before running this code, you'll add your initials to the second-to-last line.

15. Add your initials as follows: `output_name="Election2020VoterResultsBy-County ABC",`. Replace "ABC" with your initials.

16. Run the modified code.

Create a map and add the Join Results Layer

1. Enter and run the following code:

```
my_map = gis.map("USA",4)
my_map.add_layer(results)
my_map
```

2. On the map, click a county to view the pop-up.

3. In the pop-up, locate the votes_dem and votes_gop fields that you created.

4. Save your notebook.

Take the next step

You can create tasks to schedule a notebook to run automatically, with no user interaction, at a specific time or regularly. For example, if you have a notebook that downloads data from an online source that updates frequently, scheduling it to run once a day would be helpful. That way, you could take advantage of a newly updated online data source.

Check out the steps for creating a task to schedule a notebook at links.esri.com/ScheduleNotebook.

Summary

In this tutorial, you learned how to use functions available in the open-source Python module Pandas in ArcGIS Notebooks to clean up data for spatial analysis. You added content, performed a spatial analysis task, and published content to ArcGIS Online.

CHAPTER 17

Automating data integration workflows with ArcGIS Data Pipelines

Objectives

- Explore the basics of data pipelines.
- Add inputs to a data pipeline.
- Add tools and configure operations.
- Preview results.
- Run a data pipeline and add results to Map Viewer.

Introduction

ArcGIS Data Pipelines is a data integration app with a drag-and-drop interface that lets you read data from a variety of cloud stores or file-based sources, run data preparation tools to engineer the data, and output your result as a hosted feature layer. It's a simplified way to pull in data, perform data preparation workflows, and create an output—all in the cloud.

The data integration workflow is a straightforward three-step process: connect, prepare, and write output. At any point, you can preview your results to make sure you're on the right path. To start, you connect to data (sourced from ArcGIS Online, Google BigQuery, Amazon S3, and more), add a tool and configure it, define an output, and run the data pipeline.

Additionally, if the source data is updating or changing, ArcGIS Data Pipelines can be run on a recurring basis using the task scheduler, keeping your feature layers up to date.

Tutorial 17-1: Create and modify a data pipeline

In this tutorial, you'll prepare storm path and sea shipping route data to see which storms have affected the shipping routes. With that information, you can plan better for future shipping routes. First, you'll add the storm paths and shipping route dataset, then you'll add a tool to filter the storm paths to show only hurricanes. Next, you'll add a tool to filter hurricanes based on the extent of the shipping lanes. Finally, you'll add a field that indicates whether the hurricane occurred in the extent of the shipping lanes, and you'll write a feature layer as output with that information.

When you run this data pipeline, the entire operation will occur in the cloud: you won't need to download any data, and you won't generate unnecessary intermediate layers that were made at different stages of the process. You also won't need to upload the final result. Working in the cloud provides all these benefits!

Open the Data Pipelines app and create a data pipeline

1. In a browser, sign in to ArcGIS Online.
2. On the top toolbar, click the app launcher (nine dots) and click Data Pipelines.

3. In the upper right, click Create data pipeline.

   The data pipeline editor opens. The input pane opens automatically and provides a variety of locations from which to add data.

   > *Credits are consumed when the editor status indicator says "Connected." To learn more about credit consumption, click the status indicator and click the Learn More button.*

Add data inputs

Next, you'll add a layer of observed storm paths and a file of shipping lanes.

1. Under ArcGIS, click Feature layer.
2. In the item browser, click My content and click ArcGIS Online.
3. In the search box, type **owner:Top20EssentialSkillsForArcGISOnline observed storms.**

4. In the search results, click Observed Storms. At the bottom right, click Add.

   The feature layer element appears in the canvas.

ObservedStorms
Feature layer

On your own

Add another input, following a similar process. In the Inputs pane, under File, click File. In the Select a file dialog box, click Browse existing. Search ArcGIS Online, using the term **owner:Top20EssentialSkillsForArcGISOnline shipping lanes** to locate the GeoJSON item.

5. On the canvas, drag the elements to reposition them, with Shipping Lanes below ObservedStorms.

ObservedStorms
Feature layer

Shipping Lanes
File

Preview the dataset

Throughout your workflow, Data Pipelines enables you to check your progress by visualizing a preview of your data. Next, you'll preview the attributes and locations of ObservedStorms.

1. In the canvas, select the ObservedStorms layer and click the Preview button.

   The ObservedStorms attribute table lists storms from January through July 2023. In the table, under the STORMTYPE field, several types of storms are listed. Later, you'll filter these storm types and limit the dataset to only hurricanes, but first you'll preview the layer to show all the storms on a map.

2. On the left of the table, click Map Preview.

ObservedStorms
329 preview records
Map preview
11
12

The map now shows the storm paths.

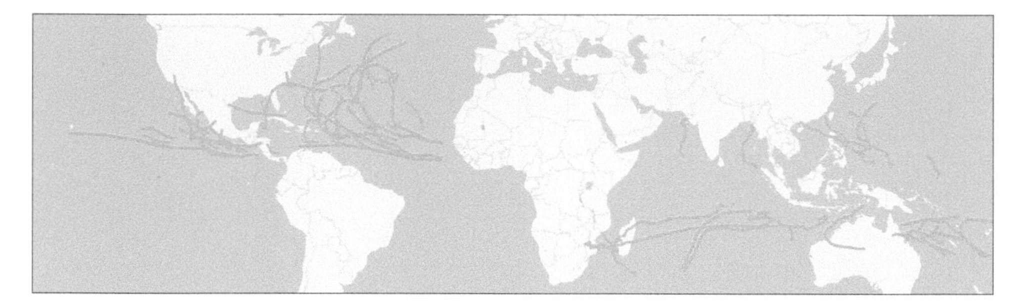

3. After you review the map, click the X in the upper right to close the preview window.

On your own

Using the same method, preview the Shipping Lanes attributes and locations. Three shipping lanes are shown. Close the preview.

Filter by attribute

You'll use a filter to show only the storms that reached hurricane force (with maximum sustained winds of 74 mph).

1. On the editor toolbar, click Tools.

2. Click Filter by attribute.

   The Filter by Attribute tool is added to the canvas.

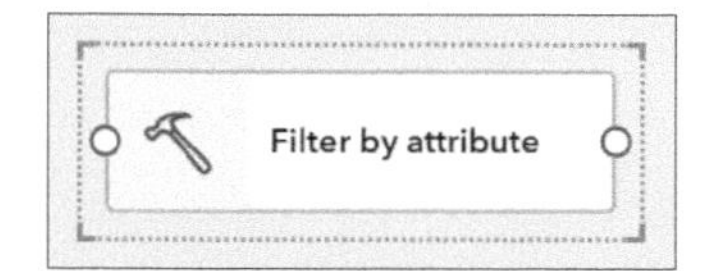

3. Drag the Filter by Attribute tool element to the right of the ObservedStorms feature layer.

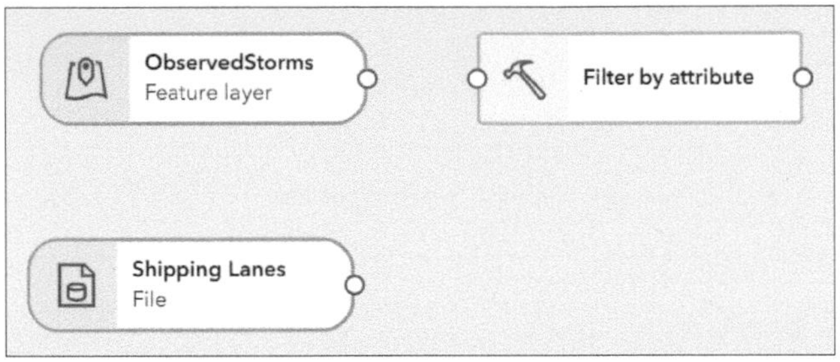

You have two options for connecting elements in the canvas and building your data pipeline. You'll try the visual method first.

4. In the canvas, click the output port (small circle) of the ObservedStorms element and drag your pointer to the input port of the Filter by Attribute tool element.

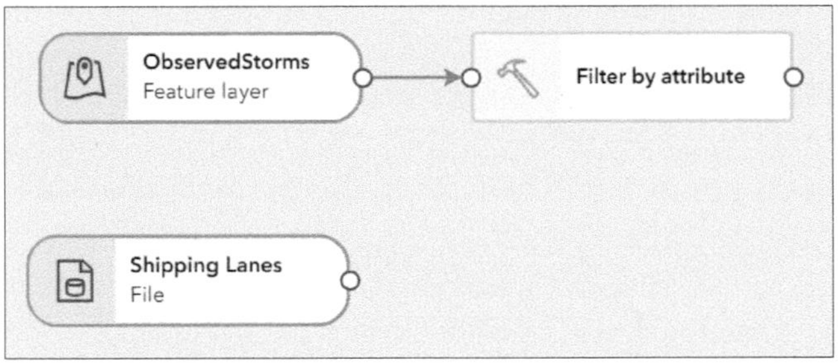

This visual method of connecting elements gives a clear picture of the workflow you're building.

5. In the Filter by attribute pane, under Filter expression, click Build new query.

6. In the Query builder window, with Expression selected, click Next.

7. Build the following query: Where all of the following are true STORMTYPE contains **Hurricane**.

> *You must type "Hurricane."*

Query builder

Show records from

ObservedStorms

Where

All of the following are true

STORMTYPE | contains | Hurricane

8. Click Add.

   This filter limits the display to only hurricane storm tracks.

   > *STORMTYPE includes five categories of hurricanes: Hurricane1, Hurricane2, and so on. By filtering for any STORMTYPE that contains the word* Hurricane, *you'll include all five categories and avoid the need to create an expression listing each hurricane category.*

On your own

Select the filter and check the map preview. Now that a filter is applied, fewer storms should be shown on the map. Click a storm track. In the pop-up, look at the STORMTYPE field to confirm that it's a hurricane. Close the preview.

Filter by extent

Now you'll add the Filter by Extent tool to limit the hurricanes to show those hurricanes that occur in the extent of the Shipping Lanes layer.

1. In the editor toolbar, click Tools and click Filter by extent.

   The Filter by Extent tool element is added to the canvas.

2. Drag the Filter by Extent tool element to the right of the Filter by Attribute tool element.

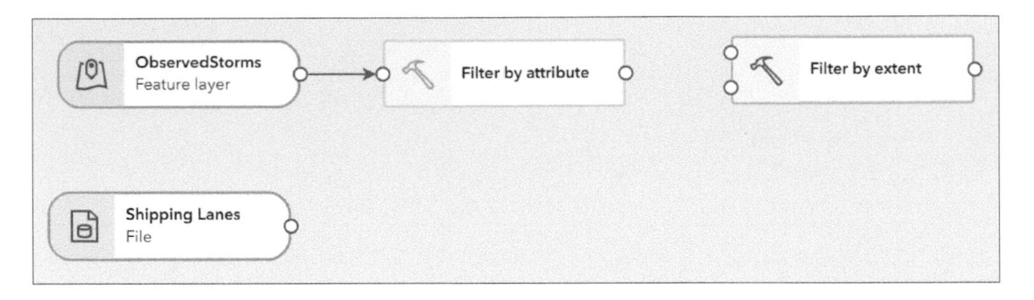

Now you'll try the second way of connecting elements in the canvas—by using the pane.

3. In the Filter by extent pane, under Input dataset, click Dataset.
4. In the Select dataset pane, click Filter by attribute.
5. In the Filter by extent pane, under Extent type, click Use the extent of another dataset.
6. Under Extent dataset, click Dataset.
7. In the Select dataset pane, click Shipping Lanes.
8. In the Filter by extent pane, confirm that the Extent geometry field has been automatically populated with the geometry field.
9. In the upper right, on the action bar, click Auto layout diagram.

   The positions of elements on the canvas are rearranged to better illustrate the processes.

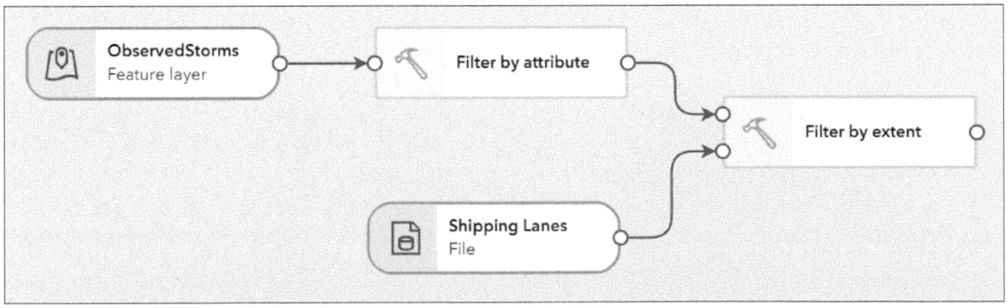

ObservedStorms
Feature layer
Filter by attribute
Filter by extent
Shipping Lanes
File

On your own

On the Filter by extent pane, click Preview. View the map preview. Confirm that the results are as expected and the storms shown are hurricanes. Verify that the storms occur in the general extent of the shipping lanes. Close the preview.

Add a field

Next, you'll add a field to your eventual output to confirm that the hurricanes occurred in the extent of the shipping lanes.

1. On the editor toolbar, click Tools and click Calculate field.
2. Drag the Calculate New Field tool element to the right of the Filter by Extent tool element.

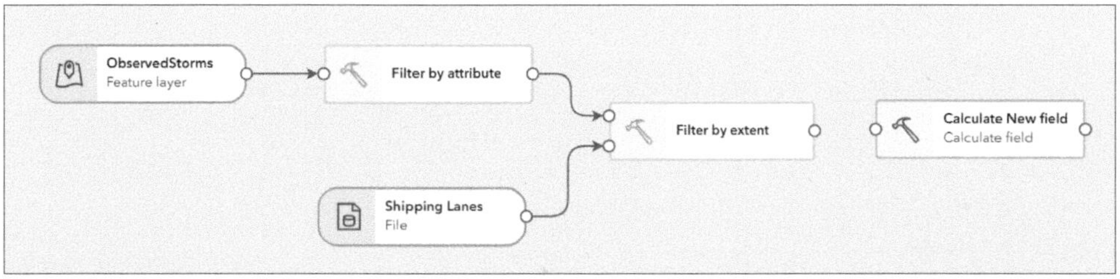

ObservedStorms
Feature layer
Filter by attribute
Filter by extent
Calculate New field
Calculate field
Shipping Lanes
File

On your own

Connect the Filter by Extent tool to the Calculate New Field tool using the visual method or the pane.

3. In the Calculate New Field pane, under Field to calculate, confirm that New field is selected.
4. Under New field name, type **HurricanesForAnalysis**.
5. Under New field type, click and select String.
6. Under Arcade expression, click Author Arcade expression.
7. In the Arcade expression builder, type **"Hurricanes impacting shipping lanes"**.
8. Click Save.

Specify the output

You've added input, filtered by attribute to show hurricanes, filtered by extent to show hurricanes in the shipping routes, and added a field to indicate what these records mean. Now you'll specify the output.

1. On the editor toolbar, click Outputs. In the Outputs pane, click Feature layer.
2. Connect the Calculate New Field element to the Create feature layer element using the visual method or the pane.
3. Click Auto layout diagram. 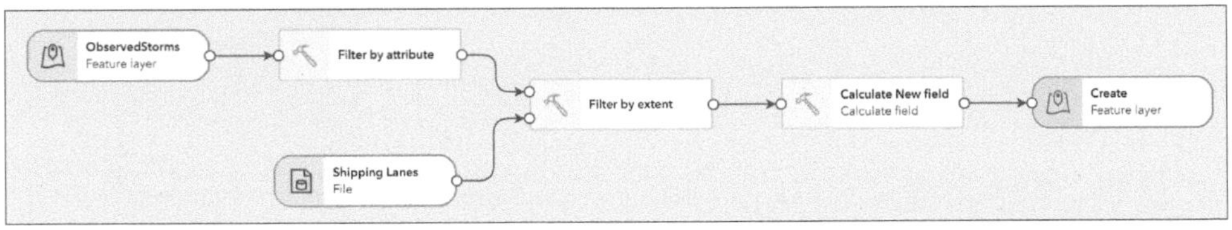
4. In the Create pane, for Output name, type **HurricanesImpactingRoutes <your initials>**.
5. For Folder, choose Top 20 Tutorial Content.

   > *If you need to create the folder, click the down arrow and choose Create new folder. Type* **Top 20 Tutorial Content**.

6. On the action bar, at the upper right, click Run.

   The Latest run details console shows the progress. The data pipeline creates the layer, ready to be used in a dashboard, app, or analysis workflow to help make important decisions.

Save your data pipeline

1. On the editor toolbar, click Save > Save as.
2. Save your data pipeline with the following settings:
    - Title: **Shipping lanes**
    - Folder: Top 20 Tutorial Content
    - Tags: **Top 20, data pipeline**
    - Summary: **A data pipeline to view shipping lanes and hurricane impacts.**
3. Click Save.

Review the result

1. On the Latest run details console, click the Output results tab.
2. Click the layer you've created.

    A new browser tab opens, showing you the item page for your new layer.
3. Close the browser tab with the data pipeline.

    > *Close the browser tab to disconnect from the app and conserve your credit consumption.*
4. On the item page, click Open in Map Viewer.

    Map Viewer shows hurricane paths that can affect shipping. Now you're ready to begin your analysis with your prepared data.

Take the next step

In Map Viewer, add the shipping routes to the map by searching for **owner:Top20EssentialSkillsForArcGISOnline shipping lanes.** To visualize your newly prepared data, use the STORMTYPE field to style the HurricanesImpactingRoutes layer and illustrate the hurricane category. Style the shipping lanes. Perform your analysis, perhaps exploring the proximity of storms to a shipping lane of interest. Save and share your map.

Summary

In this chapter, you created a data pipeline. You added two datasets, filtered by an attribute and an extent, and added and calculated a new field. Your output provided you with a layer of hurricanes that occurred in your area of interest, allowing further analysis of impacts on shipping routes.

Workflow

1. From the app launcher, click Data Pipelines.
2. Add inputs.
3. Add processing tools.
4. Configure tools to run.
5. Preview your result.
6. Specify an output.
7. Run the data pipeline.

CHAPTER 18

Joining a group and working with group content

Objectives

- Become a member of a group.
- Learn how groups are configured and how content is organized and managed.
- Create a group, invite members, and add group content.

Introduction

A group is a collection of items that are shared among members. A group allows members to work together on projects. Items in a group are often worked on and edited by members, who prepare those items for sharing to the public. Groups can also be used to keep certain items private. Group owners and managers can create a group and decide who can find it, who can join it, and who can contribute content.

In this chapter, you'll explore a group's content, edit an item that's available only to group members, and create your own group.

Tutorial 18-1: Explore the content in an existing group

In this tutorial, you'll join a group, learn how content is organized, find content of interest within the group, and edit a group item.

Sign in and find a group

If you completed chapter 1, you already joined the group (Top 20 Essential Skills for ArcGIS Online) that was set up for this book. If you didn't complete chapter 1 or aren't sure, you can complete the first few steps to see whether you're already a member. If you aren't a member, you'll have a chance to join.

1. Sign in to ArcGIS Online.
2. On the top navigation bar, click Groups.

   On the blue navigation bar, My groups is selected.

3. If the results show the group—Top 20 Essential Skills for ArcGIS Online—click View group details, skip the rest of this section, and jump to the next section ("View Group Details"). If you don't see the group, continue with the next step.
4. On the top navigation bar, click the magnifying glass.
5. In the search box, type **Top 20 Essential Skills for ArcGIS Online**, and press Enter.
6. On the left, click the Groups tab.

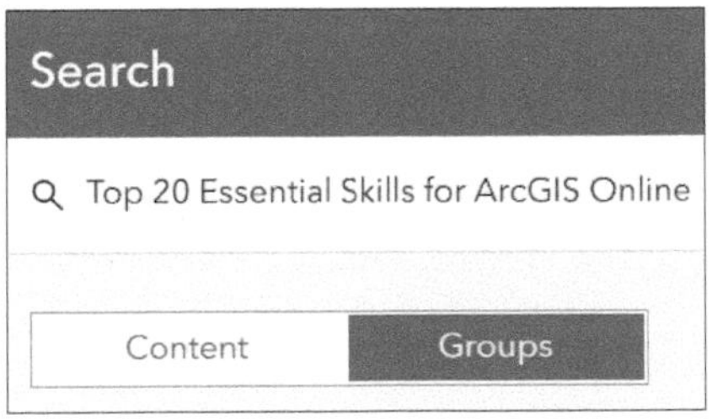

   Search

   Top 20 Essential Skills for ArcGIS Online

   Content | Groups

7. Under Filters, find the toggle button next to the name of your organization and turn it off.

   The group appears.

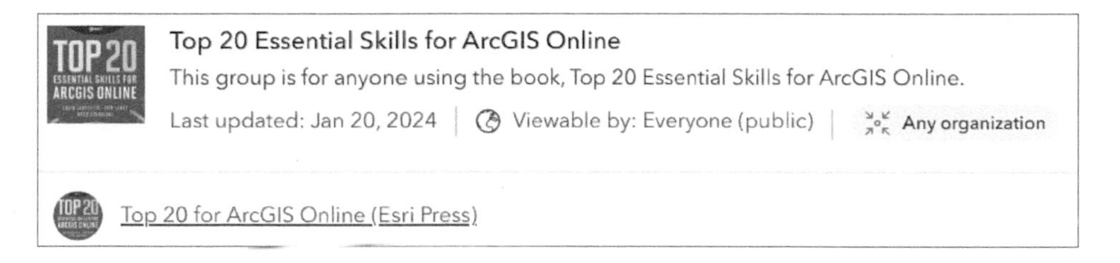

8. On the Top 20 Essential Skills for ArcGIS Online group, click View group details.

9. On the group overview page, on the right, click Join this group.

   You are now listed as a member.

   Membership

   ✓ You are a member

View group details

On the blue navigation bar, Overview is selected, showing the description of the group.

1. On the right, under Details, confirm that the group has been configured to allow anyone to view the group, but only group owners and managers can contribute content and see members.

   At the bottom of the page, group content is featured. This location can be used to highlight items that may be of interest to most group members.

Find group items

Group items are located on the Content tab.

1. On the blue navigation bar, click Content.

   On the left are filters that help you locate items of interest. A group owner or manager defines group categories to help organize items in the collection.

2. Under Group categories, click each of the categories to review the content.

Filters

- Group categories
  - Uncategorized
  - Boundaries (1)
  - Roads (1)
  - Trees (2)
  - Points of Interest (2)

3. Under Points of Interest, notice the two items listed.

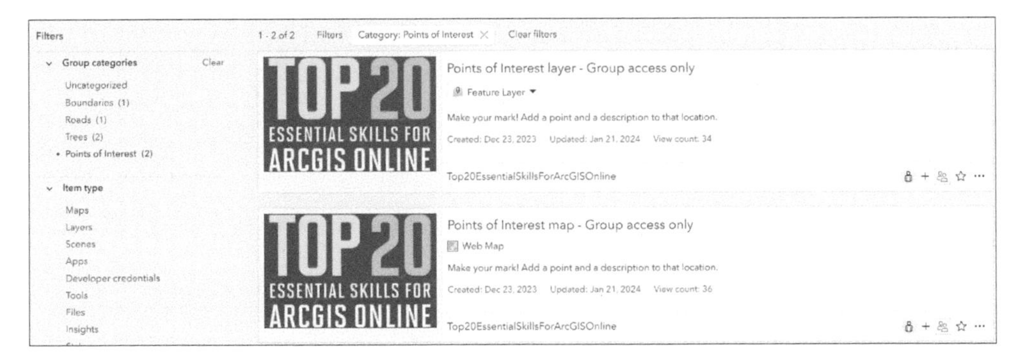

One item is a layer and the other is a map. These items are shared only to this group, so only group members have access to them.

4. At the lower right of each item, click the star to mark the items as favorites.

   You'll review your favorites later.

Edit an item in a group

This content is accessible only to members of this group, so access is somewhat limited. But this group is open to anyone who wants to join. However, group membership can be limited, for example, by restricting membership to only those who have been invited or by reviewing requests to join before allowing access to the group.

1. In the group content, in the Points of Interest category, click the thumbnail for the Points of Interest web map – Group access only.

   The map opens in Map Viewer. The layer in this map has been configured to allow editing, so you'll add a point of interest to share with others who are using this book and have joined the group.

2. On the Settings toolbar, click Edit.
3. In the Editor pane, click New Feature.

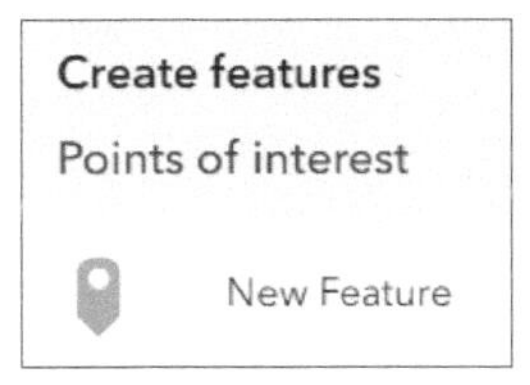

4. Zoom and pan in the map to a location that you find interesting—a place you've visited, want to visit, or think is notable.
5. Click to add a point.
6. In the Create features pane, add a description.

Create features
Settings
Description*
I'd like to visit this place someday

7. At the bottom of the pane, click Create.
8. On the Settings pane, click Edit to conclude editing, and close the Editor pane.

   Editing has been configured to allow members to add points and update their own points and descriptions. Members are also able to see points added by others. But members can edit only their own points. The configuration options allow for a variety of scenarios that facilitate collaboration but also preserve the integrity of the data.

9. Click some of the other points that have been added to view the pop-ups.

   Now you'll return to the group to review your favorites.

10. In the upper left, click the menu and click Content.

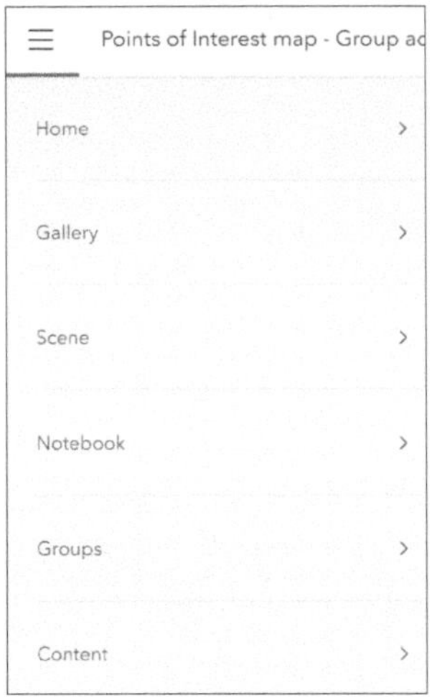

Points of Interest map - Group ac

Home

Gallery

Scene

Notebook

Groups

Content

View favorites and group content

Next, you'll review favorite content and group content.

1. On the blue navigation bar, click My favorites.

   The Points of Interest layer and map you set as favorites are listed, along with other items you may have set as favorites earlier.

2. On the blue navigation bar, click My groups.

   By default, you can see all content from the Top 20 Essential Skills for ArcGIS Online group and any other group that you're a member of.

   You have a few different ways to organize items you need for convenience, depending on how often you need to access them.

Tutorial 18-2: Create a group and configure access

In this tutorial, you'll create a group, configure access, and invite members. It's important to have a clear idea of how the group will be used and what privileges you want to grant members.

You'll set up a group for members of your organization to view content that you've decided to share. You'll configure the group so that anyone in your organization can view it but can only join if you invite them. Only you can add content to the group. You'll also send group invitations and monitor whether you receive any other group invitations.

Create a group

Creating a group takes only a few steps.

1. On the top navigation bar, click Groups.

   Any groups of which you're a member are listed.

2. In the upper left, click Create group.

3. In the Create a group window, add the following information:

   - Name: **Project collaboration <your initials>**
   - Summary: **A general group for collaboration.**
   - Tags: **Top 20, Groups, Collaboration**

On your own

Find or create a thumbnail for your group.

4. Configure the following settings:

   - Who can view this group? All organization members.
   - Who can be in this group? My organization's members only.
   - How can people join this group? By invitation.
   - Who can contribute content? Group owners and managers.

5. Click Save.

   The group is configured.

Invite others to join your group

Now that you've created a group, you'll invite members.

1. In the upper right of the group Overview page, click Invite members.

   You'll configure your invitations to allow invited members to review the invitation and confirm that they want to join the group.

2. In the Invite members window, at the bottom, uncheck the Add organization members without requiring confirmation box.

3. Find classmates or other members of your organization whom you want to invite. If needed, use the search box.

4. Next to the username, click the check box and click Invite members.

   The invitations are sent for review.

Add items to your group

Now you can share any content that you want to collaborate on.

1. In the upper right, click Add items to group.

2. Pick an item of your choice, click the check box, and click Add items.

   Hint: In addition to adding your own items, you can add public items from ArcGIS Online or ArcGIS Living Atlas. In the Add items to group window, in the upper left, click My content, change the selection to the collection you want to search, and add the items.

   Your new content appears under Recently added content.

Manage group membership and invitations (optional)

This section is optional—others may not have had time to review your group invitation yet.

You'll check whether your invitation has been accepted and members have been added to your group.

1. On the blue navigation bar, click Members.

2. Review the membership list to see whether your invitation has been accepted and a new member has been added.

3. On the top navigation bar, to the left of your profile, click the notifications button (bell).

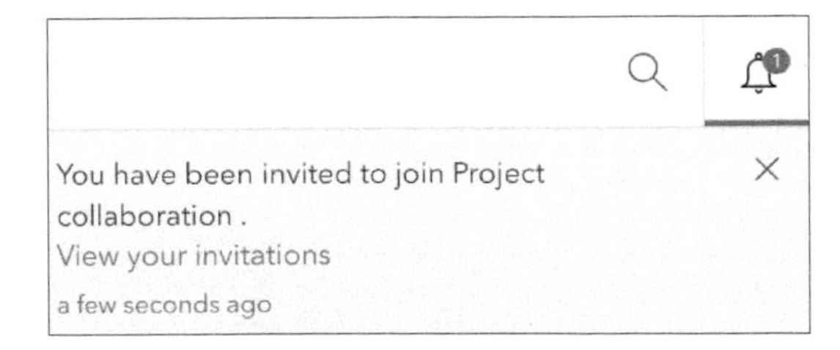

4. Click View your invitations.
5. On the group, choose Decline or Join this group.

Summary

In this chapter, you joined a group, explored group content, and edited an item that was exclusive to the group. You learned several ways that group content can be organized: by content, by category, by filtering, by group, and by setting a favorite. Finally, you configured and created your own group, added an item to your group, invited members, and confirmed whether a new member joined your group.

Workflow

1. On the top navigation bar, click Groups.
2. In the upper left, click Create group.
3. Fill out the group details.
4. Configure the group settings.
5. In the upper right of the Overview page, click invite members.
6. Determine whether you want to automatically add others or let them review the invitation.
7. Add members.
8. Click Add content, locate content of interest, and click Add items.

CHAPTER 19

Sharing an app in ArcGIS Instant Apps

Objectives

- Learn about ArcGIS Instant Apps.
- Choose a template.
- Choose a map and configure express settings.
- Add functionality from full setup mode.
- Publish an app in ArcGIS Instant Apps.

Introduction

In this chapter, you'll explore ArcGIS Instant Apps, the next generation of configurable web mapping apps. In your browser, you'll select an Instant Apps template, choose a map to showcase, configure settings, and publish your app.

Tutorial 19-1: Create and configure an app with Instant Apps

A large gallery of Instant Apps templates is available, each tailored to provide specific functionality for a need. These templates reside in a gallery, with samples that illustrate the app. If you're happy with the default configuration of the template, you can add a map and publish it in seconds, using express mode. If you want to customize the template or adjust settings, you can switch to full setup mode to add capabilities. Whichever mode you choose, using Instant Apps is an easy and straightforward workflow that requires no coding.

Open Instant Apps

1. In a browser, sign in to ArcGIS Online.

   Before you look at different options for displaying a web map, you should understand the map you want to use in the app and think about how users will interact with it. In this tutorial, you'll use the map of Ireland you made in chapter 5 and find a suitable app for it.

   > *If you have another web map you'd prefer to showcase in Instant Apps, you're encouraged to use it instead.*

2. From your contents, locate and open the Ireland map.

   This reference map shows cities, roads, and counties in Ireland. Refamiliarize yourself with the layers and design of the map and think about how users might interact with it. You want to define your goals for the final app.

3. Return to the home page. At the top of the page, click Content.

4. On the left, click Create app.

5. From the list, click Instant Apps.

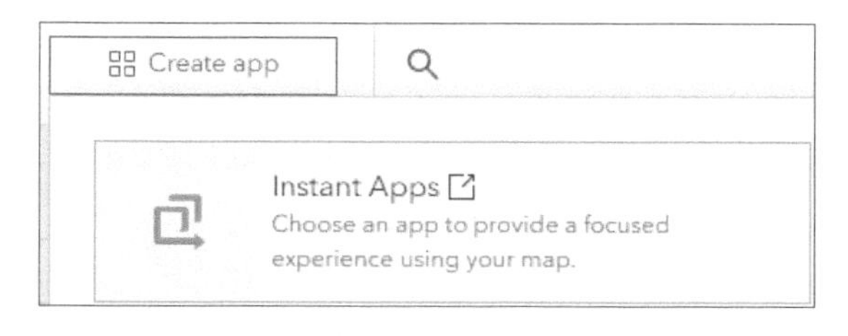

   The template gallery opens.

Choose a template

The tab on the left asks questions to help you locate an appropriate template.

1. For question 1 (What is your goal for the app?), select Showcase one or more maps with essential tools.

   The template options are filtered based on this and subsequent responses.

2. For question 2 (What do you want to display in the app?), select One web map.

3. For question 3 (Does the map include an image service?), select No.

   A final list of templates that would meet your needs is shown.

4. On the Basic template, click Sample.

A sample app using the Basic template appears.

5. Zoom and pan to explore the map. Consider whether this functionality would be appropriate for the Ireland map.

   The Basic template will display the map of Ireland in a simple layout. This template is suitable for the Ireland map.

6. In the lower right, click Choose.

7. In the pane, apply the following settings to save your app:

   - Title: **Ireland**
   - Tags: **Top 20, Ireland, Instant Apps**
   - Folder: Top 20 Tutorial Content

   > *If you haven't created the Top 20 Tutorial Content folder in an earlier tutorial, you may save the app in your root folder.*

8. Click Create app.

   The app is created.

Configure express settings

Instant Apps automatically validates that your map and data meet the app requirements. Next, you'll configure the settings.

1. In the upper left, confirm that the Express toggle is turned on by default.

   The Express setting indicates that you're viewing limited settings, which are best for quick configuration.

2. In the Express Setup pane, click Step 1.

   Next, you'll choose the Ireland map (or another map of your choice).

3. On the side panel, click Select a map or scene.

4. In the Select a map or scene window, locate the Ireland web map. Under the map, click Select map.

   > *If you're having trouble locating the web map, in the search box, type* **Ireland**.

The app configuration window updates and the map of Ireland appears.

5. On the side panel, click Next.
6. On the side panel, next to Header, click the toggle button.

   At the top of the map configuration window, the header displays the title of the app—Ireland.

   By default, a legend is included in the map. The legend is accessible from a button in the upper right of the map.

7. On the side panel, click Next.

   Now you'll configure the interactive elements.

8. Turn on the toggle buttons for Disable scroll and Screenshot.

   Disabling scroll will prevent the map from zooming when app users scroll. This setting should be disabled if users will be using mobile devices or if you plan to embed the map in a website. The Screenshot tool allows viewers to take a screenshot and download an image of the map. Adding these widgets makes your app more versatile.

9. On the side panel, click Next.

   Now you'll customize the theme and layout.

10. Under Select a preset theme, click the list and click Forest.

    The header updates to green. You'll rearrange the widgets so that users can easily find and use the functionality you're adding.

11. On the side panel, under Manage widget positions, click each widget and drag it to the upper-right section.

    The widgets are rearranged so that they appear in the upper right of the map.

    You've made some great improvements in the app configuration, but there's one more configuration you want to make that's not included in express mode.

Configure a setting in full setup mode

To see how full setup mode provides additional options, you must turn off express mode.

1. In the upper left, click the Express toggle button to turn it off.

2. When you see the warning about turning off express mode, click Continue.

   On the side panel, under Map, the Map area setting appears. This setting limits navigation on the map, which keeps users focused on the area of interest.

3. Click the Map area toggle button to turn it on, and click Configure.

   Map area
   Configure

4. In the Configure map area window, click the Navigation boundary toggle button.

   This setting activates an extent limit for users.

5. Use the boundary controls in the corners and on the sides to set an appropriate extent.

   > *You want users to be able to zoom out far enough to see the map content but not to navigate from the area of interest.*

6. Under Map view, click Set.
7. On the lower right, click Save.

On your own

Pan and zoom around the map to test your app's functionality. In the lower left, click Views. Click the Desktop, Portrait, and Landscape views. Confirm that all widgets work as desired.

8. At the bottom of the side panel, click Publish > Confirm.

   When publishing completes, the Share pane appears.

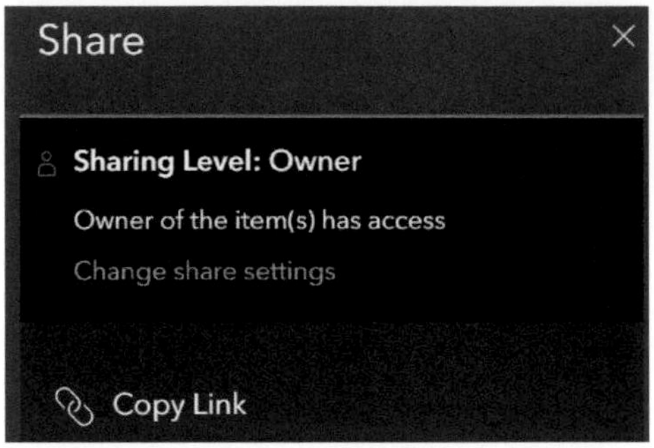

9. Click Copy Link.

   You'll test this link in a moment.

10. In the Share pane, click Change share settings.

11. Under Set sharing level, click Everyone. Click Save.

    The app has been published.

On your own

In an incognito or private browser window, paste the link to the app that you copied. Test the app. This method lets you confirm that you correctly published your content.

Take the next step

Use the skills you've just learned, try a new template using a different map. Choose a map and consider the functionality you want to include, find a suitable template, and configure settings as needed. Start with express mode but explore full setup mode, as well.

Summary

In this chapter, you learned how to choose an Instant Apps template, add a map, and configure express settings, adding functionality that lets users turn layers on and off and capture screenshots. You modified the theme and layer and switched to full setup mode to add functionality for limiting map navigation. Finally, you published your map in a professional app without having to write a single line of code.

Workflow

1. Open Instant Apps.
2. Choose a template.
3. Save the app.
4. Configure the app:
   - Choose a map.
   - Include information about the map.
   - Configure the interactive elements.
   - Customize the theme and layout.
5. Publish and share the app.

CHAPTER 20

Sharing your story

Objectives

- Craft an ArcGIS StoryMaps[SM] story.
- Add a cover and title.
- Add a sidecar and insert a map.
- Add an image to a slide.
- Create an express map.
- Configure a map action.

Introduction

ArcGIS StoryMaps is the professional storytelling tool for showcasing your geographic work in an interactive format. Using ArcGIS StoryMaps can enhance a map's impact, adding content, illustrating spatial relationships, and enhancing visual appeal.

You can surround your maps and stories with text, photos, and videos to create an interactive narrative that's easy to understand.

Tutorial 20-1: Share your work in ArcGIS StoryMaps

In this tutorial, you'll make an ArcGIS StoryMaps story to highlight the work you've done throughout this book.

> *If you haven't completed any other tutorials in this book and don't have your own content to add to the story, you can add public maps from ArcGIS Online or ArcGIS Living Atlas and modify the description of those maps as needed.*

Download content for your story

First, you'll download content to use in your story.

1. In a browser, go to **links.esri.com/Top20StoryMap**.
2. On the StoryMaps Data item, click Download to download the StoryMapsData.zip file to a convenient location on your computer.
3. Unzip the file.

   > *You'll be uploading these files and won't need them after that, so don't worry about keeping them. You can delete them from your computer after you're done.*

4. In your browser, sign in to ArcGIS Online.
5. In the upper right, on the top navigation bar, click the app launcher.

   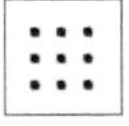

6. Click ArcGIS StoryMaps.

   The ArcGIS StoryMaps page opens.

7. In the upper right, click New story > Start from scratch.

   + New story
   Start from scratch

   Story builder launches, which provides a template for you to add content.

Add an image and title to your story

Next, you'll add an image and title to your story.

1. In the upper right, click Add cover image or video.
2. Click Browse your files and browse to the location where you unzipped the file you downloaded. Add the Cover file.

   Hint: If you prefer, you can drag the file into the window.

3. Click Add.
4. At the top of the window, click Design.

   The Design pane opens on the right.

5. Turn on the Navigation toggle button.

   Once you start adding text and style it as Heading 1, that text will automatically be added to the top of the story and function as a link to different sections in your story.

6. Under Theme, click Ridgeline.

   This preconfigured theme applies an attractive style to your story.

7. Close the Design pane.
8. In the story, click Story title and type **Web Map Portfolio**.
9. For subtitle, type **A collection of web maps that showcase a wide range of capabilities.**
10. For Name, type your name.

Add text to your story

Next, you'll add text that introduces your maps and describes your work.

1. Under the heading section, next to Tell your story, click to add a content block (plus sign).

   The block palette appears.

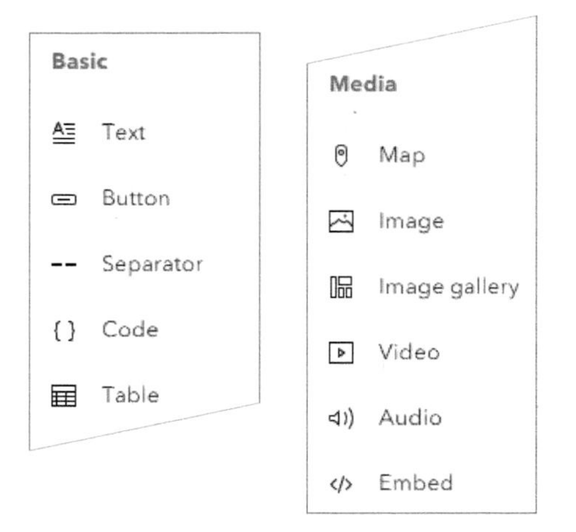

   You'll use this palette to insert content in your story. You can add images, videos, maps, code, and more.

2. Click Text.
3. Type **This collection of web maps demonstrates my understanding of essential ArcGIS Online skills.**

On your own

Below the text you added, add another text content block. Type **Colorado Rental Housing.** Select the text and style it as Heading 1. Scroll to see the beginning of the story. Notice that the navigation bar now includes the text "Colorado Rental Housing," which will allow quick navigation to the text location.

Add a sidecar to your story

Next, you'll insert a sidecar to contain a map.

1. Under the heading, click to add content.
2. In the block palette, under Immersive, click Sidecar.

   The Choose a layout window provides options for how your text and content can be structured.

3. Confirm that Docked is selected, and click Done.

   The sidecar is added to the story and is ready to be configured.

4. In the story builder, click Add > Add Map.
5. If you created the Smart Mapping map in chapter 2, click it to add it.

   > **Hint:** If you didn't create the Smart Mapping map earlier, you can choose a different map from another location such as ArcGIS Living Atlas.

   The map is added.

6. On the left, click the options button (gear).
7. Turn on the Legend toggle button, and then turn on Keep legend open.
8. Turn on the Overview map toggle button.

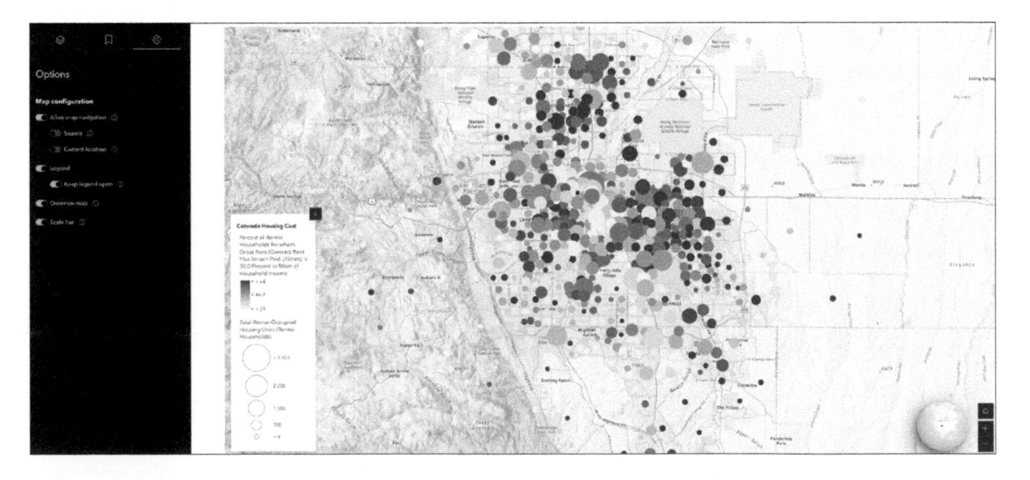

9. In the lower right, click Place map.

   The map is added to the sidecar.

10. On the left panel, click Continue your story, and type **This map shows rental housing trends near Denver, Colorado.**

Add an image to your sidecar

A sidecar consists of one or more slides that can display content. You'll add an image.

1. In the lower right of the sidecar, click New slide (plus sign).

2. In the story builder, click Add > Image or video.

3. In the Add an image or video window, browse to the folder you downloaded and add the SidecarSlide file. Click Add.

4. On the left of the panel, type **This image shows the counties that were studied.**

   Now that you have two completed slides, you'll set the transition between them.

5. Locate the list of slides.

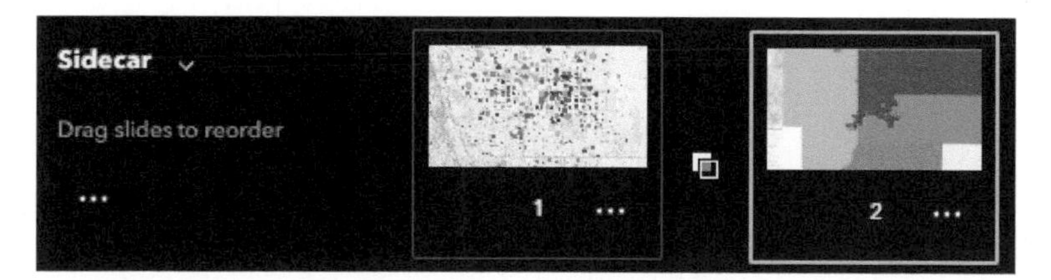

6. Click the button separating the two slides. Click Slow fade.

7. At the top of the window, click Preview.

   > *When you are warned that previewing your story will reset your undo history, click Yes, continue.*

8. Preview your story to watch the slide transition. When you have finished, at the lower right, click the X.

On your own

To practice what you just learned, add another heading, sidecar, and map. Scroll below the existing sidecar and add a heading. Add a new content block, and choose Sidecar. Choose a map to display. Include text describing the map.

Add an express map and configure map actions

You can create and add a new map even if you're in the middle of crafting your story. You don't need to rely on adding content you've already created. Next, you'll add another sidecar, configure it, and insert an express map. In that map, you'll add three places of interest. You'll add text describing the map and set map actions that let users click text that highlights a corresponding location on the map.

On your own

Under the last sidecar you made, add a text block. For the text, type **About Me.** Style the text as Heading 1.

1. Add a content block and choose Sidecar.
2. Under Choose a layout, click Floating and click Done.

   Next, you'll add an express map.

3. In the story builder, click Add > Add Map.
4. In the Add a map window, at the upper right, click New express map.

   In the map designer, you'll add three locations you want to visit.

5. On the left panel, click the options button.
6. Under Select basemap, click the default and choose a light-colored basemap—for example, the Human Geography Light basemap.

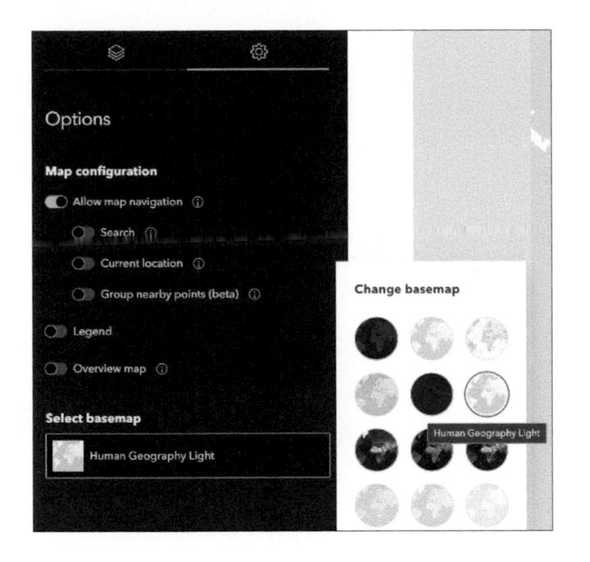

7. At the top of the window, click the Add numbered points button.

8. On the map, pan and zoom to a location you want to visit, and add it.

On your own

Click Add numbered points, and add a second and third location to the map.

9. Click Place map.

   Next, you'll add text to describe these locations.

10. In the panel, type **My top three places to visit** and press Enter.

11. Type the name of the first location, and select the text to highlight it.

12. On the menu, click Map Action.

> *This example uses Nairobi, but your location will differ.*

13. In the Configure map action window, pan and zoom to the first location. Click Save action.

On your own

Repeat these steps to add and assign map actions to the other two locations.

Preview and publish your story

1. At the top, click Preview.
2. Check your story content and functionality. In the lower right, click the preview options to see what the story will look like on different devices.
3. When you have finished previewing, in the lower right, click X.

   Now you're ready to publish your story.

4. At the top of the window, click Publish.
5. Under Share, click Private and click Everyone (public).
6. Click Publish.

> **Hint:** Update and share any items in your story that weren't previously shared publicly.

Your story is published.

Take the next step

Add maps and reorder the blocks to best present your work. Modify the cover size and theme. Refine the map descriptions to further explain the content and highlight map functionality.

Summary

In this chapter, you made a story using ArcGIS StoryMaps. Your collection of maps is now contextualized and presented in an attractive and interactive package that's easy to embed in a website or on social media.

Workflow

1. Plan your story.
2. Add a cover and title.
3. Choose a theme.
4. Write an introductory paragraph.
5. From the block palette, add basic, media, or interactive content.
6. Configure map actions.
7. Preview and publish your story.

Conclusion

Congratulations on reaching the end of the book!

Along the way, you've learned at least 20 essential skills. You can now use ArcGIS Online with confidence to make web maps, 3D scenes, notebooks, apps, and more. We hope you had fun and are excited to apply your new skills to whatever projects come your way.

ABOUT ESRI PRESS

Esri Press is an American book publisher and part of Esri, the global leader in geographic information system (GIS) software, location intelligence, and mapping. Since 1969, Esri has supported customers with geographic science and geospatial analytics, what we call The Science of Where®. We take a geographic approach to problem-solving, brought to life by modern GIS technology, and are committed to using science and technology to build a sustainable world.

At Esri Press, our mission is to inform, inspire, and teach professionals, students, educators, and the public about GIS by developing print and digital publications. Our goal is to increase the adoption of ArcGIS and to support the vision and brand of Esri. We strive to be the leader in publishing great GIS books, and we are dedicated to improving the work and lives of our global community of users, authors, and colleagues.

Acquisitions

Stacy Krieg
Claudia Naber
Alycia Tornetta
Craig Carpenter
Jenefer Shute

Editorial

Carolyn Schatz
Mark Henry
David Oberman

Production

Monica McGregor
Victoria Roberts

Sales & Marketing

Eric Kettunen
Sasha Gallardo
Beth Bauler

Contributors

Christian Harder
Matt Artz

Business

Catherine Ortiz
Jon Carter
Jason Childs

Related titles

Top 20 Essential Skills for ArcGIS Pro

Bonnie Shrewsbury & Barry Waite

9781589487505

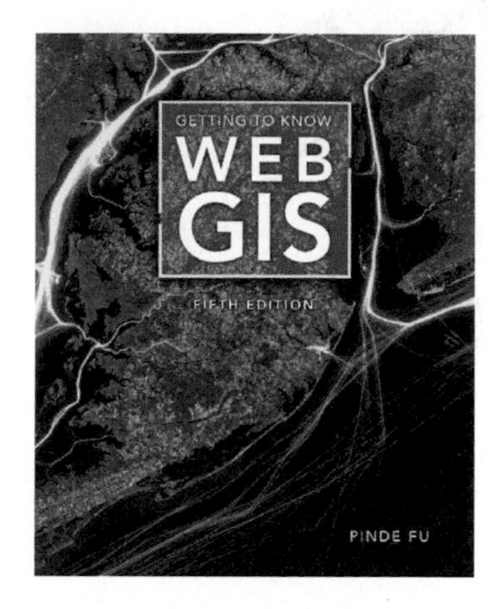

Getting to Know Web GIS, fifth edition

Pinde Fu

9781589487277

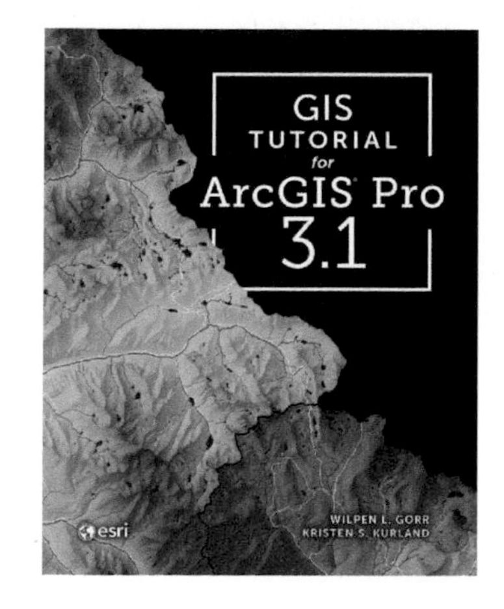

GIS Tutorial for ArcGIS Pro 3.1

Wilpen L. Gorr & Kristen S. Kurland

9781589487390

Introduction to Human Geography Using ArcGIS Online, second edition

J. Chris Carter

9781589487468

For more information about Esri Press books and resources,
or to sign up for our newsletter, visit

esripress.com.